I Have a Dream

I Have a Dream

Writings and Speeches that Changed the World

MARTIN LUTHER KING, JR.

Foreword by Coretta Scott King

Edited by
James Melvin Washington

HarperSanFrancisco
A Division of HarperCollins*Publishers*

The editor expresses gratitude to Quinton H. Dixie for his assistance in the preparation of the manuscript for this book.

FIRST EDITION

Library of Congress Cataloging-in-Publication Data

King, Martin Luther, Jr., 1929–1968.
I have a dream: writings and speeches that changed the world / Martin Luther King, Jr. : foreword by Coretta Scott King : edited by James Melvin Washington. — 1st ed.
p. cm.
Includes index.
ISBN 0-06-250552-1
1. Afro-Americans—Civil rights. 2. Civil rights movements—United States—History—20th century. 3. Nonviolence.
I. Washington, James Melvin. II. Title.
E185.97.K5A25 1992
323'.092—dc20 91-55420
CIP

07 08 HAD 50 49 48 47 46 45 44 43

This edition is printed on acid-free paper that meets the American National Standards Institute Z39.48 Standard.

Contents

Foreword

by Coretta Scott King

IN ADDITION TO DELIVERING HUNDREDS OF SPEECHES AND SERMONS DURING the American Civil Rights Movement, Martin Luther King, Jr., wrote five books and published numerous articles. With the publication of this edition of *I Have a Dream: Writings and Speeches that Changed the World,* we now have an accessible, yet representative anthology of his writings.

This collection includes many of what I consider to be my husband's most important writings and orations, including his most well-known speech, "I Have a Dream." But many other equally important works, which have been too often overlooked, are also included here.

Among them are his "Letter from a Birmingham Jail," which has been hailed as a major statement on religious responsibility in social struggle. Those who want to better understand Dr. King's development as a leader can learn much from "Pilgrimage to Nonviolence," his account of the spiritual and intellectual journey that propelled him to the forefront of the American Civil Rights Movement.

This volume also features my husband's critique of the "black power" movement ("Black Power Defined") and "A Time to Break Silence," his statement of opposition to the war in Vietnam and the destructive effects of militarism. In "The Drum Major Instinct," Dr. King offers a "new definition of excellence," a call to higher values that seems even more relevant today than when it was written. In

"Where Do We Go from Here?" he outlines his global vision for the future.

These and others of the twenty pieces in this volume capture the thought and eloquence of Martin Luther King, Jr., and together comprise a good introductory sample of his published works. It is important to remember, however, that, while he was so widely celebrated for his stirring oratory, all of the words in these pages were forged in the crucible of action. Dr. King was not merely a great speaker but a passionately committed American patriot who repeatedly put his life on the line to make real the promise of democracy.

Most of the original source materials for these and the rest of Dr. King's works can be found in the Library and Archives of the Martin Luther King, Jr., Center for Nonviolent Social Change in Atlanta, Georgia. This unique historical resource, which is used by more than 5,000 scholars and researchers every year, contains more than two million documents pertaining to the American Civil Rights Movement, including my husband's speeches, sermons, personal papers and other writings.

In his introduction and prefatory comments, James Washington has done an admirable job of putting these well-chosen selections in proper historical context. I hope this collection will encourage readers to further investigate the philosophy and methods of nonviolence that Martin Luther King, Jr., embraced and applied in the Movement and that today remain our best hope for a more just, compassionate and peaceful world.

To the Reader

To read the speeches and writings of Martin Luther King is to appreciate the courage of his moral commitment and determination to achieve social justice for his people. It is also to sense the passion of his words and to realize a terrible wrong. The America he addressed was different from the America of today. It was a nation whose racial wrongs were sanctioned by unjust laws. African Americans were singled out as prime victims and forced to accept a system designed to maintain the inferiority of their lives.

In rejecting this condition, Martin Luther King used the power of persuasion and the influence of his position to involve black Americans in realizing that they *were* somebody. By his example, he convinced black Americans that they could overcome an oppressive system. They had only to act by joining him in demonstrating for their human rights. In this exercise Martin Luther King became the most important civil rights leader in the twentieth century.

Inspired as never before, black Americans protested their segregation. As a result of massive demonstrations from the mid-1950s until the late 1960s, black Americans were to share with whites the equal use of public facilities, full participation in voting and the dignity of human rights.

CHOICE OF LEADERSHIP

In 1954 the civil rights of African Americans were more imaginary than real. Their segregation was the law of the South. This law gave a Montgomery, Alabama, bus driver the right to have a forty-two-year-old black seamstress, Rosa Parks, arrested for refusing to surrender her seat to a white passenger. When word of the December 1 incident reached the National Association for the Advancement of Colored People (NAACP), a civil rights organization, its president quickly called a meeting of local community leaders. The time had come for a bus boycott to demonstrate that black Americans were no longer willing to tolerate this mistreatment.

The one-day boycott, planned for December 5, took on a life of its own, one demanding organization and, above all, dynamic leadership. The civil rights leaders discussed the qualifications needed. Ideally, such a person would have to be young, vigorous, and educated and would also have to be courageous and an effective speaker. A minister seemed the most logical choice since the black church provided the spiritual strength that would be needed. The president of the NAACP had such a man in mind—young, well trained, and fresh to the community. His recommendation was Martin Luther King, Jr., and the others agreed.

For all his qualifications, King at first appeared a questionable choice. He was hardly a fiery freedom fighter. As the son and grandson of pastors, he was more privileged than deprived, more protected from than victimized by racism. Growing up in Atlanta, Georgia, he was a good student, entering Morehouse, a local black college, at age fifteen, and graduating first in his class from Crozer Theological, a predominantly white seminary in Pennsylvania.

Before completing his doctorate at Boston University, King went back to the South and in September of 1954 assumed the position of pastor of Montgomery's Dexter Avenue Baptist Church. Its small but influential congregation was made up of respectable professionals, the type of people King was accustomed to and comfortable with.

The only distraction to the young minister's new career was the presence of Montgomery's racism. He felt and deeply resented its existence, even before the incident calling for the bus boycott. Learning of Rosa Parks's arrest, he realized that black people had to do something about racial discrimination, and when called upon to lead the way, King readily accepted the challenge.

CHALLENGING RACIAL WRONGS

Named to head the newly formed Montgomery Improvement Association (MIA), the twenty-six-year-old pastor was now responsible for leading the December 5 boycott. Its objective was to force the city to change the law that prohibited black Americans from sitting in the front of public buses with whites. The city's white officials refused and for 381 days the boycott continued. During this time 90 percent of the black riders stayed away, and the bus company's revenues dropped by some 65 percent. Finally, the United States Supreme Court upheld Montgomery's three-judge federal district court ruling, declaring "Alabama's state and local laws requiring segregation on buses unconstitutional."

This decision sparked a new beginning in the struggle of black Americans for civil rights. By introducing nonviolent protest as an effective strategy against unjust laws in the South, Martin Luther King emerged as a national and international symbol of racial justice. As head of the Southern Christian Leadership Conference, he began a campaign of open confrontation with those upholding legal segregation.

King believed that only through a massive nonviolent assault would conditions change for black Americans. Such a nonviolent movement was launched in Montgomery, refined by student sit-ins, agitated by mass marches, and demonstrated by Freedom Rides, and finally rewarded by passage of the Voting Rights Act of 1965, which gave black Americans equal access to the polls with Southern whites.

A LASTING TRIBUTE

Under King's leadership, more African Americans than ever before were inspired to demonstrate for their constitutional rights. His speeches and writings chronicle this history. They also serve as a lasting reminder of the more just laws he helped establish. They are not to be taken for granted, nor should the moral conscience he inspired be forgotten. They are a written record of the man and of what he represented—for this nation to fulfill its commitment "to the proposition that race has no place in American life or law."

During the final year and a half of his life, King concluded that racism, poverty and the Vietnam War were interrelated and equally wrong in robbing the nation of its vitality. Looking to 1968, he began planning a Poor People's March on Washington.

On April 3, 1968, King took occasion to reminisce about the civil rights activities that had taken place since 1955 and expressed his gratitude for having played a part in this progress. "I just want to do God's will," he said. "And He's allowed me to go up to the mountain. And I've looked over. And I've seen the promised land. . . . And I'm happy tonight. I'm not worried about anything. I'm not fearing any man. Mine eyes have seen the glory of the coming of the Lord."

It was a vision for which Martin Luther King labored until his untimely death.

—Warren J. Halliburton, author and educator

Editor's Introduction

IF MARTIN LUTHER KING, JR., WERE ALIVE TODAY, I WOULD ASK HIM, "What were you thinking about as you were waiting to deliver your 'I Have a Dream' speech that August day in 1963? Were you astonished at the size of the integrated crowd—over 200,000 demonstrators gathered at the Lincoln Memorial in Washington, D.C.—who eagerly waited to hear what you had to say? Were you wondering at how it was possible for a black man who was only 34 years old to be pushed to the center stage of American history so rapidly? Were you nervous about revealing your precious dream to a whole nation?" The cowardly malice of assassination denies us the privilege of asking such questions of the great American prophet of the twentieth century.

Since we cannot query Dr. King, I would like to offer a historian's autopsy of a strange corpse called "Jim Crow" who was Martin King's Goliath. This creature, a caricature created by discrimination, was an autocratic giant in a nation that professed to be democratic. The roots of his genealogy and the influence of his progeny, however, spread far beyond the borders of the United States. This was not apparent to many Americans. But sometimes crises, especially international ones, force those who are asleep to wake from their slumber.

Many people were unaware that the world was on the brink of social and economic collapse in 1929 when Martin Luther King, Jr., was born into the family of a middle class African American minister on Auburn Avenue in Atlanta, Georgia. Ten months after King's birth, the Western world succeeded in mismanaging itself into a mammoth economic depression. But economic depressions within the African American community were always more severe and certainly more frequent than in other communities. These people were constantly migrating to different parts of the United States in search of personal and economic dignity. The benefits of American urbanization and industrialization had bypassed most black people.

Because of his family's middle-class standing, Martin Luther King, Jr., was more fortunate than most other black children. Yet economic well-being could not change the social stigma against African American people in a country where the white majority belittled and discriminated against people of color. Segregationists spent untold amounts of public funds to remind black people of their subordinate place in American society. While tired black adults yearned and struggled for a better day for their children, these children dreamed of a different future. Martin Luther King, Jr.'s, dream world, like the rest of humanity, began its steady march toward consciousness on his birthday, January 15, 1929. The absurd dissonance of the drumbeats of racism, however, unfairly increased this natural tempo for black youngsters.

When Martin was only four months old, a group of white vigilantes lynched a black teenager in Alamo, Tennessee, on May 29. In Princess Anne, Maryland, blacks and whites clashed on July 14. Less than a month later, on August 11, blacks and whites fought each other on East 100th Street in New York City. Twelve days later, twenty persons in Baltimore, Maryland, were injured because of another racial battle. The specific events that led to these clashes were seldom clear. But these emotional upheavals did reflect the existence of deep-seated resentment. Mutual racial antagonism of this sort had become a perverse custom since the advent of modern colonialism in the sixteenth century. A brief review of the relation between international events and the bitter racial conflicts in the years of Dr. King's maturation will help set the context for the global importance of his dream, as well as show why and how it evolved.

Europeans advanced the idea and practice of forming colonies in distant lands as an economical way of meeting their own material needs and desires. They defended this practice by gradually formulating what we now call "paternalism." Paternalism is the belief that most forms of social relations are parental in nature. Bolstered by Christian notions of charity, many Europeans came to believe that someone must always take care of those less fortunate than themselves. The unfortunate ones were to repay the fortunate ones for their "kindness" by either willingly or unwillingly submitting to forced labor. Western Europeans used their advanced technological knowledge to subjugate ancient people of different races and cultures in the name of "developing" them.

These modern rulers came to see themselves as the caretakers of their newly acquired subjects. This peculiar largess was not reserved for blacks only. Women, Indians, and non-Europeans in general were also victimized by what some pejoratively call "white male dominance." A sense of unfair class distinction among the less privileged white males became quite evident in the English Civil Wars of the mid-seventeenth century. As class lines became more difficult to cross, some leaders among the white colonialists in certain parts of the Western Hemisphere, most notably in the North American British colonies, began to denounce and resist this form of social and political inequality. Some of them, such as the American patriot Tom Paine, mused that all people must be viewed as equals in order for the colonists to escape the pitfalls of hypocrisy.

This problem of social inequality has plagued American public life since the new republic declared its independence from Great Britain in 1776. Although the particular problem of racial bigotry is as old as human tribalism, the career of this peculiar social and economic institution entered a new phase during the eighteenth century. It was then that doctrines of white supremacy acquired the status of laws of nature and were tragically institutionalized in the social mores and political life of the new republics spawned by the American and French revolutions. Even though the Constitution of the United States mandated that American involvement in the international African slave trade cease in 1808, the trade itself continued in the Western Hemisphere well into the late nineteenth century.

Limiting the international supply of slaves, however, did not constrict the domestic slave trade. The slave population increased naturally through childbirth. According to the United States Census of

1860, out of a population of nearly 32 million, there were well over four million slaves in the United States who had developed a transformed but still distinctly African culture. The nation reached the ominous conclusion that a civil war had to be fought to solidify national unity and determine the political and cultural destiny of the "Negro."

As President Abraham Lincoln declared in his Second Inaugural Address, delivered March 4, 1865, "One-eighth of the whole population were colored slaves, not distributed generally over the Union, but localized in the southern part of it. These slaves constituted a peculiar and powerful interest. All knew that this interest was somehow the cause of the war." In the midst of that war, which was fought between 1861 and 1865, President Lincoln issued an executive order, commonly called "The Emancipation Proclamation."

Although Lincoln had announced in his First Inaugural Address (March 4, 1861) that he had no intention, "directly or indirectly, to interfere with the institution of slavery in the States where it exists," several factors, including the weariness of the prolonged conflict, encouraged him to acquiesce. The Emancipation Proclamation linked the quest for national unity with the liberation of the slave population. In both his Gettysburg Address (November 19, 1863) and his Second Inaugural Address, he offered brooding but eloquent confessions that social equality in America is more of a dream than a reality.

By assassinating Lincoln in 1865, John Wilkes Booth succeeded in not only murdering the greatest American president of the nineteenth century, but also in almost mortally wounding Lincoln's hope for a new era of racial justice. Between 1865 and 1877, the period often called "Reconstruction," the nation briefly allowed African Americans to become partners in its political institutions. But social superstitions, often sanctioned by prominent intellectuals and religious leaders, underwrote a renewal of state laws, which brought an end to this historic experiment in interracial democracy.

Fear of a distended national government led to pernicious reassertions of the importance of local government. This perspective was later called "states' rights." But since local governments are more vulnerable to the dictates of regional prejudice, they led the way in the post-Reconstruction effort to disfranchise the liberated slaves. Legalized racial discrimination dominated American life between

1877 and 1954. Although Congress passed a Civil Rights Bill in 1875, the Supreme Court declared it unconstitutional in 1883. Segregationists enacted laws and enhanced social practices, commonly called "Jim Crow," in order to nurture segregated education, as well as housing discrimination against African, Jewish, and even many Catholic Americans. A hysterical fear of interracial sexual relations and interracial marriages abounded during this period, and can still be found in the late twentieth century. Hatred of African American people, and also of Jews, encouraged the cruel vigilantism that fostered the tacit, if not open, public approval of random white terrorism in the form of raping, lynching and burning.

White racism could also be quite petty. A system of American apartheid evolved. White businesses created separate accommodations for black people. But "Negroes" had to pay the same, if not more, for less. Separate parks, hospitals, cemeteries, public transportation, water fountains, public restrooms, libraries, hotels, restaurants, theaters, and schools became common throughout the country. Although this practice excelled in the South, it was actually a national phenomenon. These practices, especially in regard to education, gained legal sanction in the 1896 *Plessy v. Ferguson* Supreme Court decision, which argued that public institutions for black and white people can be "separate but equal." This practice allowed government at all levels to force African Americans to support better public accommodations for white people, and inferior ones for themselves.

This crucible of racism produced a bicultural society that was inherently unequal and hypocritical. Paul Laurence Dunbar, the famous black poet, whispered in print, "We wear the mask." The Jim Crow era enshrined racism, and attempted to entomb the spirit and the minds of millions of African Americans from the end of Reconstruction in 1877 to the landmark *Brown v. Topeka Board of Education* Supreme Court decision on May 17, 1954—a span of nearly eighty years. John Hope Franklin, the renowned African American historian, estimates that between 1884 and the outbreak of World War I, well over 3,600 African Americans were lynched. These figures do not include white rapes of black males and females, or the countless absurd indignities that the African American populace encountered as they tried to survive in a society where they were unwanted.

The irony of the sordid history of American racism is that it happened in a country that espoused the principles of democracy. Martin Luther King, Jr., was born in the midst of passionate attempts to remind the nation that racism is a subversion of its democratic and religious values. But the tide of history seemed to be against such dreamers. Scholars such as W. E. B. DuBois had argued for many years before King's birth that racism is an international phenomenon. World War II and the Nazi Holocaust offered macabre illustrations of some malevolent consequences of the simplistic acceptance of racism as a relatively harmless local custom.

By 1935, when Martin Luther King, Jr., was ready to enter the first grade at Atlanta's David T. Howard Elementary School, Adolf Hitler's "big lie" about the superiority of the Aryan race had wedded itself to a morbid fear of communism. The democratic Weimar Republic in Germany had been overthrown and the Third Reich had been established by force and chicanery. A tribalistic spirit grasped Germany. As economic chaos descended upon nearly every nation on the planet, totalitarian governments outlawed political imagination. These regimes developed rapidly in Germany, Italy, the Soviet Union and Japan. Although the degree of oppression against African Americans was certainly not on the scale of Nazi Germany's destruction of its Jewish population, its psychological effect was similar. Many Americans found the resurgence of the Ku Klux Klan, as well as the fascist and racist impulse among various right-wing Christian leaders, such as Father Charles E. Couglin and later the Reverend Gerald L. K. Smith, to be an embarrassment to American democracy.

Many decent people of all races, religions and political persuasions joined forces to resist. On March 12, 1930, a few weeks after Martin Luther King, Jr., celebrated his first birthday, those who dreamed of a world free of racism were greatly inspired by the determination of Mahatma Gandhi of India. He began his great "Salt March" to the Gulf of Gambay in protest against the audacity of the British Empire to tax Indian salt. Gandhi taught his people to use nonviolent direct action against the militarily powerful British Empire. He had already experimented with this approach to social change while a young attorney in South Africa.

The African American community had its own legacy of resistance against white supremacy, however. This tradition had taken five major forms: cultural opposition, institutional and organizational resistance, litigation, boycotts and mass protests. Although the inventiveness and spontaneity of African American culture is now legendary, the struggle to gain recognition of their unique contribution to Western culture has been arduous. The Harlem Renaissance of the 1920s is the benchmark of the success of that struggle in academic, literary and artistic circles. The pollen of this beautiful blossom helped renew the spiritual and intellectual souls of civil rights workers throughout the twentieth century.

From their inception in the late eighteenth century such African American institutions as churches and fraternal orders struggled against white supremacy. They became the mainstay of all subsequent efforts. Opposition often led to the formation of ad hoc organizations such as the National Negro Conventions, which usually met on an annual basis between 1830 and the 1850s. There were many other similar organizations throughout the nineteenth century. But none of them were as successful as the National Association for the Advancement of Colored People (NAACP) and the National Urban League (NUL), which were organized in 1909 and 1911 respectively.

The NAACP's successful litigation and its journal, *Crisis,* enhanced its image as the most effective African American civil rights organization. With the brilliant juridical strategies of African American attorneys such as Charles Houston and Thurgood Marshall, the NAACP's crowning achievement was the *Brown v. Topeka* Supreme Court decision of 1954. The diplomacy, social scientific data gathering and urbane sophistication of the NUL contributed greatly to dismantling and challenging glib generalizations about African Americans. Although both groups had extensive local branches, they were unsuccessful in mobilizing the black masses.

African American religious movements had far more successful experiences in tapping the collective imagination of black people. In his book, *Garveyism as a Religious Movement,* Randall K. Burkett offers a powerful argument for viewing Marcus Garvey's Universal Negro Improvement Association as a religious movement. In the 1920s Garvey galvanized a nascent black disdain for the white majority's cultural and political denigration of African people. But the federal government successfully prosecuted Garvey for mail fraud.

This did little to minimize black racial pride. The movement took other forms. For example, several of Garvey's compatriots were black clergy such as George Alexander McGuire, William Yancey Bell and Junius Caesar Austin, pastor of Chicago's influential Pilgrim Baptist Church. In fact in 1929 Austin organized a disciplined boycott movement against merchants that sold to black people, but would not employ them. This "Jobs for Negroes" Movement spread to Cleveland and New York.

The economic woes following the stock market crash of October 1929 encouraged many Americans, and especially African ones, to abandon the "Grand Ole Party" of Abraham Lincoln and turn to the New Deal program of the Democrat President Franklin Delano Roosevelt. One of the most important acts of Roosevelt's presidency was to sign Executive Order 8802, which established a temporary Fair Employment Practices Commission in 1941. A. Philip Randolph, president of the Brotherhood of Sleeping Car Workers, threatened to lead a march on Washington if Roosevelt did not do something to curtail racial discrimination in a war industry that was rapidly expanding to meet Western military needs after the outbreak of World War II.

Urban racial riots during the war, and especially in 1946, a year after the Allied forces defeated Hitler, underscored the sense of disgust that many African Americans felt at the hypocrisy of a nation that could loudly denounce Nazi racism, but would tolerate lynching, job discrimination and gross economic deprivation based on race. Ten years after the riots, Martin Luther King, Jr., a Ph.D. from Boston University, and a young pastor of a prominent black congregation in Montgomery, Alabama, led the most historic black mass movement of the twentieth century. Some of his most famous speeches and writings, included in this volume, outline Dr. King's dream of a world where democracy and freedom are cherished. But he was assassinated on April 4, 1968. Can a dream be murdered? The poet Langston Hughes apparently believed that dreams are deferred, not murdered. He exposes some of the consequences of disrupting other people's dreams in his poem, "Lenox Avenue Mural":

> What happens to a dream deferred?
> Does it dry up
> like a raisin in the sun?

Or fester like a sore—
And then run?
Does it stink like rotten meat?
Or crust and sugar over—
like a syrupy sweet?
Maybe it sags
like a heavy load.
Or does it explode?

The speeches and writings in this anthology offer an important profile of how a great prophet thinks in the midst of a whirlwind. Martin Luther King, Jr., provided disciplined calmness, reason and hope in the face of death threats and ominous signs of disillusionment within the African American community. He stood for a world free of bigotry and brimming with faith, hope, love and justice. He dared to dream of a better day in the midst of the nightmare that surrounded him. He dared to believe and sacrifice his life for a future that some believe we are beginning to occupy.

Chronology*

1929	*January 15.* Martin Luther King, Jr., is born to Reverend and Mrs. Martin Luther King, Sr. (the former Alberta Christine Williams), in Atlanta, Georgia.
1935–1944	King attends David T. Howard Elementary School, Atlanta University Laboratory School, and Booker T. Washington High School. He passes the entrance examination to Morehouse College (Atlanta) without graduating from high school.
1947	King is licensed to preach and becomes assistant to his father, who is pastor of the Ebenezer Baptist Church, Atlanta.
1948	*February 25.* King is ordained to the Baptist ministry.
	June. King graduates from Morehouse College with a B.A. degree in sociology.
	September. King enters Crozer Theological Seminary, Chester, Pennsylvania. After hearing Dr. A. J. Muste and

*Adapted from *Martin Luther King, Jr.: A Documentary . . . Montgomery to Memphis,* edited by Flip Schulke, New York: W. W. Norton & Company, 1976, pp. 19–21.

Dr. Mordecai W. Johnson preach on the life and teachings of Mahatma Gandhi, he begins to study Gandhi seriously.

1951 *June.* King graduates from Crozer with a B.D. degree.

1953 *June 18.* King marries Coretta Scott in Marion, Alabama.

1954 *May 17.* The Supreme Court of the United States rules unanimously in *Brown v. Board of Education* that racial segregation in public schools is unconstitutional.

October 31. King is installed by Reverend Martin Luther King, Sr., as the twentieth pastor of the Dexter Avenue Church, Montgomery, Alabama.

1955 *June 5.* King receives a Ph.D. degree in Systematic Theology from Boston University.

November 17. The Kings' first child, Yolanda Denise, is born in Montgomery.

December 1. Mrs. Rosa Parks, a forty-two-year-old Montgomery seamstress, refuses to relinquish her bus seat to a white man, and is arrested.

December 5. The first day of the bus boycott. The trial of Mrs. Parks. A meeting of movement leaders is held. Dr. King is unanimously elected president of an organization named the Montgomery Improvement Association, a name proposed by Reverend Ralph Abernathy.

December 10. The Montgomery Bus Company suspends service in black neighborhoods.

1956 *January 26.* Dr. King is arrested on a charge of traveling thirty miles an hour in a twenty-five-mile-an-hour zone in Montgomery. He is released on his own recognizance.

January 30. A bomb is thrown onto the porch of Dr. King's Montgomery home. Mrs. King and Mrs. Roscoe

Williams, wife of a church member, are in the house with baby Yolanda Denise; no one is injured.

February 2. A suit is filed in federal district court asking that Montgomery's travel segregation laws be declared unconstitutional.

February 21. Dr. King is indicted with other figures in the Montgomery bus boycott on the charge of being party to a conspiracy to hinder and prevent the operation of business without "just or legal cause."

June 4. A United States district court rules that racial segregation on city bus lines is unconstitutional.

June 27. Dr. King is the guest speaker at the annual National Association for the Advancement of Colored People (NAACP) convention in San Francisco.

November 13. The United States Supreme Court affirms the decision of the three-judge district court in declaring unconstitutional Alabama's state and local laws requiring segregation on buses.

December 20. Federal injunctions prohibiting segregation on buses are served on city and bus company officials in Montgomery. Injunctions are also served on state officials.

December 21. Montgomery buses are integrated.

1957 *January 27.* An unexploded bomb is discovered on Dr. and Mrs. King's front porch.

January 10–11. The Southern Christian Leadership Conference (SCLC) is formed at the Ebenezer Baptist Church, Atlanta. Dr. King is elected its president.

February 18. *Time* magazine puts Dr. King on its cover.

September. President Dwight D. Eisenhower federalizes the Arkansas National Guard to escort nine Negro students to an all-white high school in Little Rock, Arkansas.

September 9. The first civil rights act since Reconstruction is passed by Congress, creating the Civil Rights Commission and the Civil Rights Division of the Department of Justice.

1958 *September 17.* Dr. King's book *Stride Toward Freedom: The Montgomery Story* is published.

September 20. Dr. King is stabbed in the chest by Mrs. Izola Curry, forty-two, who is subsequently alleged to be mentally deranged. The stabbing occurs in the heart of Harlem while Dr. King is autographing his recently published book. His condition is said to be serious but not critical.

1959 *February 3–March 10.* Dr. and Mrs. King spend a month in India studying Gandhi's techniques of nonviolence, as guests of Prime Minister Nehru.

1960 *January 24.* The King family moves to Atlanta. Dr. King becomes co-pastor, with his father, of the Ebenezer Baptist Church.

February 1. A lunch counter sit-in to desegregate eating facilities is held by students in Greensboro, North Carolina.

April 15. The Student Non-Violent Coordinating Committee (SNCC) is founded to coordinate student protest at Shaw University, Raleigh, North Carolina, on a temporary basis. (It is to become a permanent organization in October, 1960.) Dr. King and James Lawson are the keynote speakers at the Shaw University founding.

October 19. Dr. King is arrested at an Atlanta sit-in and is jailed on a charge of violating the state's trespass law.

October 22–27. The Atlanta charges are dropped. All jailed demonstrators are released except for Dr. King, who is ordered held on a charge of violating a probated sentence in a traffic arrest case. He is transferred to the DeKalb County Jail in Decatur, Georgia, and is then transferred to the Reidsville State Prison. He is released from the Reidsville State Prison on a $2,000 bond.

1961 *May 4.* The first group of Freedom Riders, intent on integrating interstate buses, leaves Washington, D.C., by Greyhound bus. The group, organized by the Congress for Racial Equality (CORE), leaves shortly after the Supreme Court has outlawed segregation in interstate transportation terminals. The bus is burned outside of Anniston, Alabama, on May 14. A mob beats the Riders upon their arrival in Birmingham. The Riders are arrested in Jackson, Mississippi, and spend forty to sixty days in Parchman Penitentiary.

1962 *September 20.* James Meredith makes his first attempt to enroll at the University of Mississippi. He is actually enrolled by Supreme Court order and is escorted onto the Oxford, Mississippi, campus by U.S. marshals on October 1.

1963 *March–April.* Sit-in demonstrations are held in Birmingham to protest segregation of eating facilities. Dr. King is arrested during a demonstration.

April 16. Dr. King writes the "Letter from a Birmingham Jail" while imprisoned for demonstrating.

May 3, 4, 5. Eugene ("Bull") Connor, director of public safety of Birmingham, orders the use of police dogs and fire hoses upon the marching protesters (young adults and children).

May 20. The Supreme Court of the United States rules Birmingham's segregation ordinances unconstitutional.

June 1. Governor George C. Wallace tries to stop the court-ordered integration of the University of Alabama by "standing in the schoolhouse door" and personally refusing entrance to black students and Justice Department officials. President John F. Kennedy then federalizes the Alabama National Guard, and Governor Wallace removes himself from blocking the entrance of the Negro students.

August 28. The March on Washington, the first large integrated protest march, is held in Washington, D.C. Dr. King and other civil rights leaders meet with President John F. Kennedy in the White House, and afterwards Dr. King delivers his "I Have a Dream" speech on the steps of the Lincoln Memorial.

September. Dr. King's book *Strength to Love* is published.

1964 *May–June.* Dr. King joins other SCLC workers in demonstrations for the integration of public accommodations in St. Augustine, Florida. He is jailed.

June. Dr. King's book *Why We Can't Wait* is published.

July 2. Dr. King attends the signing of the Public Accommodations Bill, part of the Civil Rights Act of 1964, by President Lyndon B. Johnson in the White House.

August 4. The bodies of civil rights workers James Chaney, Andrew Goodman and Michael Schwerner are discovered by FBI agents buried near the town of Philadelphia, Mississippi. Neshoba County Sheriff Rainey and his deputy, Cecil Price, are allegedly implicated in the murders.

December 10. Dr. King receives the Nobel Peace Prize in Oslo, Norway.

1965 *March 7.* A group of marching demonstrators (from SNCC and SCLC) led by SCLC's Hosea Williams are beaten while attempting to march across the Edmund Pettus Bridge on their planned march to Montgomery, Alabama, from Selma, Alabama, by state highway patrolmen under the direction of Al Lingo, and sheriff's deputies under the leadership of Jim Clark. An order by Governor Wallace had prohibited the march.

March 9. Unitarian minister James Reeb is beaten by four white segregationists in Selma and dies two days later.

March 15. President Johnson addresses the nation and Congress. He describes the Voting Rights Bill he will submit to Congress in two days and uses the slogan of the civil rights movement, "We Shall Overcome."

March 16. Black and white demonstrators are beaten by sheriff's deputies and police on horseback in Montgomery.

March 21–25. Over three thousand protest marchers leave Selma for a march to Montgomery, protected by federal troops. They are joined along the way by a total of twenty-five thousand marchers. Upon reaching the capitol building they hear an address by Dr. King.

March 25. Mrs. Viola Liuzzo, wife of a Detroit Teamsters Union business agent, is shot and killed while driving a carload of marchers back to Selma.

August–December. In Alabama, SCLC spearheads voter registration campaigns in Greene, Wilcox and Eutaw counties, and in the cities of Montgomery and Birmingham.

August 6. The 1965 Voting Rights Act is signed by President Johnson.

1966 *March 25.* The Supreme Court of the United States rules any poll tax unconstitutional.

May 16. An antiwar statement by Dr. King is read at a large Washington rally to protest the war in Vietnam. Dr. King agrees to serve as co-chair of Clergy and Laymen Concerned about Vietnam.

June 6. James Meredith is shot soon after beginning his 220-mile "March Against Fear" from Memphis, Tennessee, to Jackson, Mississippi.

June. Stokely Carmichael and Willie Ricks (SNCC) use the slogan "Black Power" in public for the first time, before reporters in Greenwood, Mississippi.

1967 *January.* Dr. King writes his book *Where Do We Go from Here?*

July 6. The Justice Department reports that more than 50 percent of all eligible black voters are registered in Mississippi, Georgia, Alabama, Louisiana, and South Carolina.

November 27. Dr. King announces the formation by SCLC of a Poor People's Campaign, with the aim of representing the problems of poor blacks and whites.

1968 *February 12.* Sanitation workers strike in Memphis, Tennessee.

March 28. Dr. King leads six thousand protesters on a march through downtown Memphis in support of striking sanitation workers. Disorders break out during which black youths loot stores. One sixteen-year-old is killed, fifty persons are injured.

April 3. Dr. King's last speech, entitled "I've Been to the Mountaintop," is delivered in Memphis.

April 4. Dr. King is assassinated by a sniper as he stands talking on the balcony of his second-floor room at the Lorraine Motel in Memphis. He dies in St. Joseph's Hospital from a gunshot wound to the neck. James Earl Ray is later captured and convicted of the murder.

PART I

The Dream Bursts Forth

(1956–1959)

1
Our Struggle

(1956)

Participation by African American churches in American public life surged in the 1940s and 1950s throughout the United States, but especially in the American South. These churches often formed both formal and informal alliances with organizations such as the National Association for the Advancement of Colored People (NAACP) and the National Urban League (NUL) in order to resist Jim Crow. For example, the Reverend Oliver Brown, pastor of the St. Catherine's African Methodist Episcopal Church, allowed his daughter to participate in the NAACP's litigation of the historic *Brown v. Topeka, Kansas Board of Education* Supreme Court decision of May 17, 1954.

But rather than initiate change, the religious community was often a respondent to crises originating elsewhere. On December 1, 1955, Mrs. Rosa Parks, a forty-two-year-old seamstress, refused to vacate her seat on a public bus so that a white man, as Alabama state law required, could have her seat. The bus driver called the police who immediately arrested Mrs. Parks, a well-known community activist. Four days later, on December 5, African Americans began a bus boycott after an evening rally was held. The black community unanimously elected the Reverend Dr. Martin Luther King, Jr., as the first president of the Montgomery (Alabama) Improvement Association. This nonviolent organization signaled a new alliance between black institutions such as churches, fraternal orders, professional associations of teachers and doctors, and mass protests. A strong belief in social responsibility and consciousness had permeated a new generation of church leaders who sought to wed religious convictions with the need to change various hypocrisies of American public life. There were many forerunners of this effort. Dr. King himself indicated that he learned much from prior nonviolent resisters such as the Reverend Theodore J. Jemison of Baton Rouge, Louisiana, who had led a successful boycott against that city's public bus system as early as 1953, and from the earlier attempts of the Reverend Vernon Johns, who preceded King as pastor of the Dexter Avenue Baptist Church in Montgomery.

After the African American community of Montgomery, Alabama, proudly walked rather than ride the buses for over eleven months, the United States Supreme Court affirmed a lower court's decision to declare unconstitutional Alabama's segregationist laws that required racial separation on buses. It took another month to force state and local officials to respect the Supreme Court's interpretation of the law. The buses were finally integrated on December 21, 1956.

The following essay, which was published in the religious journal *Liberation,* is a summary of Dr. King's book, *Stride Toward Freedom*. It offers a passionate analysis and defense of the Montgomery Bus Boycott. This succinct essay was one of the first summaries of the motivations and objectives that eventually led to the formation of the Southern Christian Leadership Conference in 1957.

THE SEGREGATION OF NEGROES, WITH ITS INEVITABLE DISCRIMINATION, HAS thrived on elements of inferiority present in the masses of both white and Negro people. Through forced separation from our African culture, through slavery, poverty, and deprivation, many black men lost self-respect.

In their relations with Negroes, white people discovered that they had rejected the very center of their own ethical professions. They could not face the triumph of their lesser instincts and simultaneously have peace within. And so, to gain it, they rationalized—insisting that the unfortunate Negro, being less than human, deserved and even enjoyed second-class status.

They argued that his inferior social, economic and political position was good for him. He was incapable of advancing beyond a fixed position and would therefore be happier if encouraged not to attempt the impossible. He is subjugated by a superior people with an advanced way of life. The "master race" will be able to civilize him to a limited degree, if only he will be true to his inferior nature and stay in his place.

White men soon came to forget that the southern social culture and all its institutions had been organized to perpetuate this rationalization. They observed a caste system and quickly were conditioned to believe that its social results, which they had created, actually reflected the Negro's innate and true nature.

In time many Negroes lost faith in themselves and came to believe that perhaps they really were what they had been told they were—something less than men. So long as they were prepared to accept this role, racial peace could be maintained. It was an uneasy peace in which the Negro was forced to accept patiently injustice, insult, injury and exploitation.

Gradually the Negro masses in the South began to reevaluate themselves—a process that was to change the nature of the Negro community and doom the social patterns of the South. We discovered that we had never really smothered our self-respect and that we could not be at one with ourselves without asserting it. From this point on, the South's terrible peace was rapidly undermined by the Negro's new and courageous thinking and his ever-increasing readiness to organize and to act. Conflict and violence were coming to the surface as the white South desperately clung to its old patterns. The extreme tension in race relations in the South today is explained in part by the revolutionary change in the Negro's evaluation of himself and of his destiny and by his determination to struggle for justice. *We Negroes have replaced self-pity with self-respect and self-depreciation with dignity.*

When Mrs. Rosa Parks, the quiet seamstress whose arrest precipitated the nonviolent protest in Montgomery, was asked why she had refused to move to the rear of a bus, she said: "It was a matter of dignity; I could not have faced myself and my people if I had moved."

THE NEW NEGRO

Many of the Negroes who joined the protest did not expect it to succeed. When asked why, they usually gave one of three answers: "I didn't expect Negroes to stick to it," or, "I never thought we Negroes had the nerve," or, "I thought the pressure from the white folks would kill it before it got started."

In other words, our nonviolent protest in Montgomery is important because it is demonstrating to the Negro, North and South, that many of the stereotypes he has held about himself and other Negroes are not valid. Montgomery has broken the spell and is ushering in concrete manifestations of the thinking and action of the new Negro.

We now know that:

We Can Stick Together

In Montgomery, forty-two thousand of us have refused to ride the city's segregated buses since December 5. Some walk as many as fourteen miles a day.

Our Leaders Do Not Have to Sell Out

Many of us have been indicted, arrested, and "mugged." Every Monday and Thursday night we stand before the Negro population at the prayer meetings and repeat: "It is an honor to face jail for a just cause."

Threats and Violence Do Not Necessarily Intimidate Those Who Are Sufficiently Aroused and Nonviolent

The bombing of two of our homes has made us more resolute. When a handbill was circulated at a White Citizens Council meeting stating that Negroes should be "abolished" by "guns, bows and arrows, sling shots and knives," we responded with even greater determination.

Our Church Is Becoming Militant

Twenty-four ministers were arrested in Montgomery. Each has said publicly that he stands prepared to be arrested again. Even upper-class Negroes who reject the "come to Jesus" gospel are now convinced that the church has no alternative but to provide the nonviolent dynamics for social change in the midst of conflict. The thirty thousand dollars used for the car pool, which transports over twenty thousand Negro workers, school children and housewives, has been raised in the churches. The churches have become the dispatch centers where the people gather to wait for rides.

We Believe in Ourselves

In Montgomery we walk in a new way. We hold our heads in a new way. Even the Negro reporters who converged on Montgomery have a new attitude. One tired reporter, asked at a luncheon in Birmingham to say a few words about Montgomery, stood up, thought for a moment, and uttered one sentence: "Montgomery has made me proud to be a Negro."

Economics Is Part of Our Struggle

We are aware that Montgomery's white businessmen have tried to "talk sense" to the bus company and the city commissioners. We have observed that small Negro shops are thriving as Negroes find it inconvenient to walk downtown to the white stores. We have been getting more polite treatment in the white shops since the protest began. We have a new respect for the proper use of our dollar.

We Have Discovered a New and Powerful Weapon—Nonviolent Resistance

Although law is an important factor in bringing about social change, there are certain conditions in which the very effort to adhere to new legal decisions creates tension and provokes violence. We had hoped to see demonstrated a method that would enable us to continue our struggle while coping with the violence it aroused. Now we see the answer: face violence if necessary, but refuse to return violence. If we respect those who oppose us, they may achieve a new understanding of the human relations involved.

We Now Know That the Southern Negro Has Come of Age, Politically and Morally

Montgomery has demonstrated that we will not run from the struggle, and will support the battle for equality. The attitude of many young Negroes a few years ago was reflected in the common expression, "I'd rather be a lamp post in Harlem than Governor of Alabama." Now the idea expressed in our churches, schools, pool rooms, restaurants and homes is: "Brother, stay here and fight nonviolently. 'Cause if you don't let them make you mad, you can win." The official slogan of the Montgomery Improvement Association is "Justice without Violence."

THE ISSUES IN MONTGOMERY

The leaders of the old order in Montgomery are not prepared to negotiate a settlement. This is not because of the conditions we have set for returning to the buses. The basic question of segregation in intrastate travel is already before the courts. Meanwhile we ask only for what in Atlanta, Mobile, Charleston and most other cities of the South is considered the southern pattern. We seek the right, under segregation, to seat ourselves from the rear forward on a first come, first served basis. In addition, we ask for courtesy and the hiring of some Negro bus drivers on predominantly Negro routes.

A prominent judge of Tuscaloosa was asked if he felt there was any connection between Autherine Lucy's effort to enter the University of Alabama and the Montgomery nonviolent protest. He replied, "Autherine is just one unfortunate girl who doesn't know what she is doing, but in Montgomery it looks like all the niggers have gone crazy."

Later the judge is reported to have explained that "of course the good niggers had undoubtedly been riled up by outsiders, Communists and agitators." It is apparent that at this historic moment most of the elements of the white South are not prepared to believe that "our Negroes could of themselves act like this."

MISCALCULATION OF THE WHITE LEADERS

Because the mayor and city authorities cannot admit to themselves that we have changed, every move they have made has inadvertently increased the protest and united the Negro community.

[1955]

Dec. 1 They arrested Mrs. Parks, one of the most respected Negro women in Montgomery.

Dec. 3 They attempted to intimidate the Negro population by publishing a report in the daily paper that certain Negroes were calling for a boycott of the buses. They thereby informed the thirty thousand Negro readers of the planned protest.

Dec. 5 They found Mrs. Parks guilty and fined her fourteen dollars. This action increased the number of those who joined the boycott.

Dec. 5 They arrested a Negro college student for "intimidating passengers." Actually, he was helping an elderly woman cross the street. This mistake solidified the college students' support of the protest.

Two policemen on motorcycles followed each bus on its rounds through the Negro community. This attempt at psychological coercion further increased the number of Negroes who joined the protest.

In a news telecast at 6:00 P.M., a mass meeting planned for that evening was announced. Although we had expected only five hundred people at the meeting, over five thousand attended.

Dec. 6 They began to intimidate Negro taxi drivers. This led to the setting up of a car pool and a resolution to extend indefinitely our protest, which had originally been called for one day only.

Dec. 7 They began to harass Negro motorists. This encouraged the Negro middle class to join the struggle.

Dec. 8 The lawyer for the bus company said, "We have no intention of hiring Negro drivers now or in the foreseeable future." To us this meant never. The slogan then became, "Stay off the buses until we win."

Dec. 9 The mayor invited Negro leaders to a conference, presumably for negotiation. When we arrived, we discovered that some of the men in the room were white supremacists and members of the White Citizens Council. The mayor's attitude was made clear when he said, "Comes the first rainy day and the Negroes will be back in the buses." The next day it did rain, but the Negroes did not ride the buses.

At this point over forty-two thousand Montgomery Negroes had joined the protest. After a period of uneasy quiet, elements in the white community turned to further police intimidation and to violence.

[1956]

Jan. 26 I was arrested for traveling thirty miles per hour in a twenty-five-mile zone. This arrest occurred just two hours before a mass meeting. So, we had to hold seven mass meetings to accommodate the people.

Jan. 30 My home was bombed.

Feb. 1 The home of E. D. Nixon, one of the protest leaders and former state president of the NAACP, was bombed. This brought moral and financial support from all over the state.

Feb. 22 Eighty-nine persons, including the twenty-four ministers, were arrested for participating in the nonviolent protests.

Every attempt to end the protest by intimidation, by encouraging Negroes to inform, by force and violence, further cemented the Negro community and brought sympathy for our cause from men of good will all over the world. The great appeal for the world appears to lie in the fact that we in Montgomery have adopted the method of nonviolence. In a world in which most men attempt to defend their highest values by the accumulation of weapons of destruction, it is morally refreshing to hear five thousand Negroes in Montgomery shout "Amen" and "Halleluh" when they are exhorted to "pray for those who oppose you," or pray "Oh Lord, give us strength of body to keep walking for freedom," and conclude each mass meeting with: "Let us pray that God shall give us strength to remain nonviolent though we may face death."

THE LIBERAL DILEMMA

And death there may be. Many white men in the South see themselves as a fearful minority in an ocean of black men. They honestly believe with one side of their minds that Negroes are depraved and disease-ridden. They look upon any effort at equality as leading to "mongrelization." They are convinced that racial equality is a Communist idea and that those who ask for it are subversive. They believe that their caste system is the highest form of social organization.

The enlightened white southerner, who for years has preached gradualism, now sees that even the slow approach finally has revolutionary implications. Placing straws on a camel's back, no matter how slowly, is dangerous. This realization has immobilized the liberals and most of the white church leaders. They have no answer for dealing with or absorbing violence. They end in begging for retreat, lest "things get out of hand and lead to violence."

Writing in *Life,* William Faulkner, Nobel Prize–winning author from Mississippi, recently urged the NAACP to "stop now for a moment." That is to say, he encouraged Negroes to accept injustice, exploitation and indignity for a while longer. It is hardly a moral act to encourage others patiently to accept injustice which he himself does not endure.

In urging delay, which in this dynamic period is tantamount to retreat, Faulkner suggests that those of us who press for change now may not know that violence could break out. He says we are "dealing with a fact: the fact of emotional conditions of such fierce unanimity as to scorn the fact that it is a minority and which will go to any length and against any odds at this moment to justify and, if necessary, defend that condition and its right to it."

We southern Negroes believe that it is essential to defend the right of equality now. From this position we will not and cannot retreat. Fortunately, we are increasingly aware that we must not try to defend our position by methods that contradict the aim of brotherhood. We in Montgomery believe that the only way to press on is by adopting the philosophy and practice of nonviolent resistance.

This method permits a struggle to go on with dignity and without the need to retreat. It is a method that can absorb the violence that is inevitable in social change whenever deep-seated prejudices are challenged.

If, in pressing for justice and equality in Montgomery, we discover that those who reject equality are prepared to use violence, we must not despair, retreat, or fear. Before they make this crucial decision, they must remember: whatever they do, we will not use violence in return. We hope we can act in the struggle in such a way that they will see the error of their approach and will come to respect us. Then we can all live together in peace and equality.

The basic conflict is not really over the buses. Yet we believe that, if the method we use in dealing with equality in the buses can eliminate injustice within ourselves, we shall at the same time be attacking the basis of injustice—man's hostility to man. This can only be done when we challenge the white community to re-examine its assumptions as we are now prepared to reexamine ours.

We do not wish to triumph over the white community. That would only result in transferring those now on the bottom to the top. But, if we can live up to nonviolence in thought and deed, there will emerge an interracial society based on freedom for all.

Liberation 1 (April 1957): 3–6.

2

Facing the Challenge of a New Age

(1957)

The birth pangs of the "Freedom Movement," as the Civil Rights Movement was often called by contemporaries, resonate in this edited version of Dr. King's address before the First Annual Institute on Non-Violence and Social Change, held in Montgomery, Alabama, in December 1956. The National Association for the Advancement of Colored People (NAACP) and the Montgomery Improvement Association (MIA) had been victorious in convincing the Supreme Court to overturn Alabama segregation laws regarding public transportation. As the leader of the MIA, Dr. King sought to define the implications of this in his far-reaching interpretation of the philosophical possibilities of nonviolent direct action. King realized, however, that words were not enough. He immediately gathered his extensive contacts within the religious community, especially those among the huge National Baptist Convention, USA, Inc., to organize a Southern Christian Leadership Conference. He placed this new mass movement's possibilities in the context of the fall of Western colonialism. He used the MIA's achievement to illustrate his belief that nonviolent direct action was the most moral means that oppressed people have for achieving self-determination.

But there were reactionary forces in the nation that resisted any effort to achieve racial justice. White racists used dynamite on September 9, 1957, to destroy the new Hattie Cotton Elementary School in Nashville, Tennessee, that had allowed one black child to enroll. Ironically, on the same day, Congress passed the first civil rights bill since 1875. This bill established the United States Civil Rights Commission, as well as the Civil Rights Division of the Department of Justice. Meanwhile opposition arose against the 1954 Supreme Court decision that mandated the desegregation of American school systems "with all deliberate speed." On September 24 President Dwight Eisenhower used television and radio to announce that he had ordered federal troops to escort nine black children to Central High School in Little Rock, Arkansas.

Despite these tragic events, Dr. King, as we can see in the following address, searched for hope. He argued that the African American freedom movement had international parallels wherever European colonialism was still trying to maintain its feeble grip. By March 6, 1957, the African nation of Ghana acquired its independence from British colonial rule. In King's first national address, which was given before the Prayer Pilgrimage on May 17, he stated that these developments were the work of divine providence: "We proudly proclaim that three-fourths of the peoples of the world are colored. We have the privilege of noticing in our generation the great drama of freedom and independence as it unfolds in Asia and Africa. All these things are in line with the unfolding work of providence."

THOSE OF US WHO LIVE IN THE TWENTIETH CENTURY ARE PRIVILEGED TO live in one of the most momentous periods of human history. It is an exciting age filled with hope. It is an age in which a new social order is being born. We stand today between two worlds—the dying old and the emerging new.

Now I am aware of the fact that there are those who would contend that we live in the most ghastly period of human history. They would argue that the rhythmic beat of the deep rumblings of discontent from Asia, the uprisings in Africa, the nationalistic longings of Egypt, the roaring cannons from Hungary, and the racial tensions of America are all indicative of the deep and tragic midnight which encompasses our civilization. They would argue that we are retrogressing instead of progressing. But far from representing retrogression and tragic meaninglessness, the present tensions represent the necessary pains that accompany the birth of anything new. Long ago the Greek philosopher Heraclitus argued that justice emerges from the strife of opposites, and Hegel, in modern philosophy, preached a doctrine of growth through struggle. It is both historically and biologically true that there can be no birth and growth without birth and growing pains. Whenever there is the emergence of the new we confront the recalcitrance of the old. So the tensions which we witness in the world today are indicative of the fact that a new world order is being born and an old order is passing away.

We are all familiar with the old order that is passing away. We have lived with it for many years. We have seen it in its international aspect, in the form of colonialism and imperialism. There are approximately two billion four hundred million (2,400,000,000) people in this world, and the vast majority of these people are colored—about one billion six hundred million (1,600,000,000) of the people of the world are colored. Fifty years ago, or even twenty-five years ago, most of these one billion six hundred million people lived under the yoke of some foreign power. We could turn our eyes to China and see there six hundred million men and women under the pressing yoke of British, Dutch, and French rule. We could turn our eyes to Indonesia and see a hundred million men and women under the domination of the Dutch. We could turn to India and Pakistan and notice four hundred million brown men and women under the pressing yoke of the British. We could turn our eyes to Africa and notice there two hundred million black men and women under the pressing yoke of the British, the Dutch and the French. For years all of these people were dominated politically, exploited economically, segregated and humiliated.

But there comes a time when people get tired. There comes a time when people get tired of being trampled over by the iron feet of oppression. There comes a time when people get tired of being plunged across the abyss of exploitation where they experience the bleakness of nagging despair. There comes a time when people get tired of being pushed out of the glittering sunlight of life's July and left standing in the piercing chill of an Alpine November. So in the midst of their tiredness these people decided to rise up and protest against injustice. As a result of their protest more than one billion three hundred million (1,300,000,000) of the colored peoples of the world are free today. They have their own governments, their own economic systems, and their own educational systems. They have broken loose from the Egypt of colonialism and imperialism, and they are now moving through the wilderness of adjustment toward the promised land of cultural integration. As they look back they see the old order of colonialism and imperialism passing away and the new order of freedom and justice coming into being.

We have also seen the old order in our own nation, in the form of segregation and discrimination. We know something of the long history of this old order in America. It had its beginning in the year 1619 when the first Negro slaves landed on the shores of this nation. They were brought here from the soils of Africa. And unlike the Pilgrim Fathers who landed at Plymouth a year later, they were brought here against their wills. Throughout slavery the Negro was treated in a very inhuman fashion. He was a thing to be used, not a person to be respected. He was merely a depersonalized cog in a vast plantation machine. The famous Dred Scott Decision of 1857 well illustrates the status of the Negro during slavery. In this decision the Supreme Court of the United States said, in substance, that the Negro is not a citizen of the United States; he is merely property subject to the dictates of his owner. Then came 1896. It was in this year that the Supreme Court of this nation, through the *Plessy v. Ferguson* decision, established the doctrine of separate-but-equal as the law of the land. Through this decision segregation gained legal and moral sanction. The end result of the Plessy doctrine was that it led to a strict enforcement of the "separate," with hardly the slightest attempt to abide by the "equal." So the Plessy doctrine ended up making for tragic inequalities and ungodly exploitation.

Living under these conditions, many Negroes came to the point of losing faith in themselves. They came to feel that perhaps they were less than human. The great tragedy of physical slavery was that it led to mental slavery. So long as the Negro maintained this subservient attitude and accepted this "place" assigned to him, a sort of racial peace existed. But it was an uneasy peace in which the Negro was forced patiently to accept insult, injustice and exploitation. It was a negative peace. True peace is not merely the absence of some negative force—tension, confusion, or war; it is the presence of some positive force—justice, good will and brotherhood. And so the peace which existed between the races was a negative peace devoid of any positive and lasting quality.

Then something happened to the Negro. Circumstances made it necessary for him to travel more. His rural plantation background was gradually being supplanted by migration to urban and industrial communities. His economic life was gradually rising to decisive proportions. His cultural life was gradually rising through the steady decline of crippling illiteracy. All of these factors conjoined to cause the Negro to take a new look at himself. Negro masses began to reevaluate themselves. The Negro came to feel that he was somebody. His religion revealed to him that God loves all of His children, and that every man, from a bass black to a treble white, is significant on God's keyboard. So he could now cry out with the eloquent poet:

> Fleecy locks and black complexion
> Cannot forfeit nature's claim.
> Skin may differ, but affection
> Dwells in black and white the same.
> And were I so tall as to reach the pole
> Or to grasp the ocean at a span,
> I must be measured by my soul.
> The mind is the standard of the man.

With this new self-respect and new sense of dignity on the part of the Negro, the South's negative peace was rapidly undermined. And so the tension which we are witnessing in race relations today can be explained, in part, by the revolutionary change in the Negro's evaluation of himself, and his determination to struggle and sacrifice until the walls of segregation have finally been crushed by the battering rams of surging justice.

Along with the emergence of a "New Negro," with a new sense of dignity and destiny, came that memorable decision of May 17, 1954. In this decision the Supreme Court of this nation unanimously affirmed that the old Plessy doctrine must go. This decision came as a legal and sociological death blow to an evil that had occupied the throne of American life for several decades. It affirmed in no uncertain terms that separate facilities are inherently unequal and that to segregate a child because of his race is to deny him equal protection of the law. With the coming of this great decision we could gradually see the old order of segregation and discrimination passing away, and the new order of freedom and justice coming into being. Let nobody fool you, all of the loud noises that you hear today from the legislative halls of the South in terms of "interposition" and "nullification," and of outlawing the NAACP are merely the death groans from a dying system. The old order is passing away, and the new order is coming into being. We are witnessing in our day the birth of a new age, with a new structure of freedom and justice.

Now as we face the fact of this new, emerging world, we must face the responsibilities that come along with it. A new age brings with it new challenges. Let us consider some of the challenges of this new age.

First, we are challenged to rise above the narrow confines of our individualistic concerns to the broader concerns of all humanity. The new world is a world of geographical togetherness. This means that no individual or nation can live alone. We must all learn to live together, or we will be forced to die together. This new world of geographical togetherness has been brought about, to a great extent, by man's scientific and technological genius. Man through his scientific genius has been able to dwarf distance and place time in chains; he has been able to carve highways through the stratosphere. And so it is possible today to eat breakfast in New York City and dinner in Paris, France. Bob Hope has described this new jet age in which we live. It is an age in which we will be able to get a nonstop flight from Los Angeles, California, to New York City, and if by chance we develop hiccups on taking off, we will "hic" in Los Angeles and "cup" in New York City. It is an age in which one will be able to leave Tokyo on Sunday morning and, because of time difference, arrive in Seattle, Washington, on the preceding Saturday night. When your friends meet you at the airport in Seattle inquiring when you left Tokyo, you will have to say, "I left tomorrow." This, in a very humorous sense, says to us that our world is geographically one. Now we are faced with the challenge of making it spiritually one. Through our scientific genius we have made of the world a neighborhood; now through our moral and spiritual genius we must make of it a brotherhood. We are all involved in the single process. Whatever affects one directly affects all indirectly. We are all links in the great chain of humanity. This is what John Donne meant when he said years ago:

> No man is an island, entire of it selfe; every man is a piece of the Continent, a part of the maine; if a clod bee washed away by the Sea, Europe is the lesse, as well as if a Promontorie were, as well as if a Mannor of thy friends or of thine owne were; any mans' death diminishes me, because I am involved in Mankinde; And therefore never send to know for whom the bell tolls; it tolls for thee.

A second challenge that the new age brings to each of us is that of achieving excellency in our various fields of endeavor. In the new age many doors will be opening to us that were not opened in the past, and the great challenge which we confront is to be prepared to enter these doors as they open. Ralph Waldo Emerson said in an essay back in 1871:

> If a man can write a better book, or preach a better sermon, or make a better mouse trap than his neighbor, even if he builds his house in the woods the world will make a beaten path to his door.

In the new age we will be forced to compete with people of all races and nationalities. Therefore, we cannot aim merely to be good Negro teachers, good Negro doctors, good Negro ministers, good Negro skilled laborers. We must set out to do a good job, irrespective of race, and do it so well that nobody could do it better.

Whatever your life's work is, do it well. Even if it does not fall in the category of one of the so-called big professions, do it well. As one college president said, "A man should do his job so well that the living, the dead, and the unborn could do it no better." If it falls your lot to be a street sweeper, sweep streets like Michelangelo painted pictures, like Shakespeare wrote poetry, like Beethoven composed music; sweep streets so well that all the host of Heaven and earth will have to pause and say, "Here lived a great street sweeper, who swept his job well." As Douglas Mallock says:

If you can't be a pine on the top of the hill
Be a scrub in the valley—but be
The best little scrub by the side of the hill,
Be a bush if you can't be a tree.

If you can't be a highway just be a trail
If you can't be the sun be a star;
It isn't by size that you win or fail—
Be the best of whatever you are.

A third challenge that stands before us is that of entering the new age with understanding good will. This simply means that the Christian virtues of love, mercy and forgiveness should stand at the center of our lives. There is the danger that those of us who have lived so long under the yoke of oppression, those of us who have been exploited and trampled over, those of us who have had to stand amid the tragic midnight of injustice and indignities will enter the new age with hate and bitterness. But if we retaliate with hate and bitterness, the new age will be nothing but a duplication of the old age. We must blot out the hate and injustice of the old age with the love and justice of the new. This is why I believe so firmly in nonviolence. Violence never solves problems. It only creates new and more complicated ones. If we succumb to the temptation of using violence in our struggle for justice, unborn generations will be the recipients of a long and desolate night of bitterness, and our chief legacy to the future will be an endless reign of meaningless chaos.

We have before us the glorious opportunity to inject a new dimension of love into the veins of our civilization. There is still a voice crying out in terms that echo across the generations, saying: Love your enemies, bless them that curse you, pray for them that despitefully use you, that you may be the children of your Father which is in Heaven.

This love might well be the salvation of our civilization. This is why I am so impressed with our motto for the week, "Freedom and Justice through Love." Not through violence; not through hate; no, not even through boycotts; but through love. It is true that as we struggle for freedom in America we will have to boycott at times. But we must remember as we boycott that a boycott is not an end within itself; it is merely a means to awaken a sense of shame within the oppressor and challenge his false sense of superiority. But the end is reconciliation; the end is redemption; the end is the creation of the beloved community. It is this type of spirit and this type of love that can transform opposers into friends. It is this type of understanding good will that will transform the deep gloom of the old age into the exuberant gladness of the new age. It is this love which will bring about miracles in the hearts of men.

Now I realize that in talking so much about love it is very easy to become sentimental. There is the danger that our talk about love will merely be empty words devoid of any practical and true meaning. But when I say love those who oppose you I am not speaking of love in a sentimental or affectionate sense. It would be nonsense to urge men to love their oppressors in an affectionate sense. When I refer to love at this point I mean understanding good will. The Greek language comes to our aid at this point. The Greek language has three words for love. First it speaks of love in terms of *eros*. Plato used this word quite frequently in his dialogues. *Eros* is a type of esthetic love. Now it has come to mean a sort of romantic love. I guess Shakespeare was thinking in terms of *eros* when he said:

> Love is not love
> which alters when it alteration finds,
> or bends with the remover to remove.
> O no, it is an ever fixed mark
> that looks on tempests and is never shaken.
> It is a star to every wandering bark. . . .

This is *eros*. And then the Greek talks about *philia*. *Philia* is a sort of intimate affectionateness between personal friends. It is a sort of reciprocal love. On this level a person loves because he is loved. Then the Greek language comes out with another word which is the highest level of love. It speaks of it in terms of *agape*. *Agape* means nothing sentimental or basically affectionate. It means understanding, redeeming good will for all men. It is an overflowing love which seeks nothing in return. It is the love of God working in the lives of men. When we rise to love on the *agape* level we love men not because we like them, not because their attitudes and ways appeal to us, but because God loves us. Here we rise to the position of loving the person who does the evil deed while hating the deed that the person does. With this type of love and understanding good will we will be able to stand amid the radiant glow of the new age with dignity and discipline. Yes, the new age is coming. It is coming mighty fast.

Now the fact that this new age is emerging reveals something basic about the universe. It tells us something about the core and heartbeat of the cosmos. It reminds us that the universe is on the side of justice. It says to those who struggle for justice, "You do not struggle alone, but God struggles with you." This belief that God is on the side of truth and justice comes down to us from the long tradition of our Christian faith. There is something at the very center of our faith which reminds us that Good Friday may occupy the throne for a day, but ultimately it must give way to the triumphant beat of the drums of Easter. Evil may so shape events that Caesar will occupy a palace and Christ a cross, but one day that same Christ will rise up and split history into A.D. and B.C., so that even the life of Caesar must be dated by His name. There is something in this universe that justifies Carlyle in saying, "No lie can live forever." There is something in this universe which justifies William Cullen Bryant in saying, "Truth crushed to earth will rise again." There is something in this universe that justifies James Russell Lowell in saying:

> Truth forever on the scaffold
> Wrong forever on the throne
> Yet that scaffold sways the future
> And behind the dim unknown stands God
> Within the shadows keeping watch above his
>   own.

And so here in Montgomery, after more than eleven long months, we can walk and never get weary, because we know there is a great camp meeting in the promised land of freedom and justice.

Before closing I must correct what might be a false impression. I am afraid that if I close at this point many will go away misinterpreting my whole message. I have talked about the new age which is fastly coming into being. I have talked about the fact that God is working in history to bring about this new age. There is the danger, therefore, that after hearing all of this you will go away with the impression that we can go home, sit down, and do nothing, waiting for the coming of the inevitable. You will somehow feel that this new age will roll in on the wheels of inevitability, so there is nothing to do but wait on it. If you get that impression you are the victims of an illusion wrapped in superficiality. We must speed up the coming of the inevitable.

Now it is true, if I may speak figuratively, that Old Man Segregation is on his deathbed. But history has proven that social systems have a great last-minute breathing power, and the guardians of a-status quo are always on hand with their oxygen tents to keep the old order alive. Segregation is still a fact in America. We still confront it in the South in its glaring and conspicuous forms. We still confront it in the North in its hidden and subtle forms. But if democracy is to live, segregation must die. Segregation is a glaring evil. It is utterly unchristian. It relegates the segregated to the status of a thing rather than elevates him to the status of a person. Segregation is nothing but slavery covered up with certain niceties of complexity. Segregation is a blatant denial of the unity which we all have in Christ Jesus.

So we must continue the struggle against segregation in order to speed up the coming of the inevitable. We must continue to gain the ballot. This is one of the basic keys to the solution of our problem. Until we gain political power through possession of the ballot we will be convenient tools of unscrupulous politicians. We must face the appalling fact that we have been betrayed by both the Democratic and Republican parties. The Democrats have betrayed us by capitulating to the whims and caprices of the southern dixiecrats. The Republicans have betrayed us by capitulating to the blatant hypocrisy of right-wing reactionary northerners. This coalition of southern Democrats and northern right-wing Republicans defeats every proposed bill on civil rights. Until we gain the ballot and place proper public officials in office this condition will continue to exist. In communities where we confront difficulties in gaining the ballot, we must use all legal and moral means to remove these difficulties.

We must continue to struggle through legalism and legislation. There are those who contend that integration can come only through education, for no other reason than that morals cannot be legislated. I choose, however, to be dialectical at this point. It is neither education nor legislation; it is both legislation and education. I quite agree that it is impossible to change a man's internal feelings merely through law. But this really is not the intention of the law. The law does not seek to change one's internal feelings; it seeks rather to control the external effects of those internal feelings. For instance, the law cannot make a man love—religion and education must do that—but it can control his efforts to lynch. So in order to control the external effects of prejudiced internal feelings, we must continue to struggle through legislation.

Another thing that we must do in pressing on for integration is to invest our finances in the cause of freedom. Freedom has always been an expensive thing. History is a fit testimony to the fact that freedom is rarely gained without sacrifice and self-denial. So we must donate large sums of money to the cause of freedom. We can no longer complain that we do not have the money. Statistics reveal that the economic life of the Negro is rising to decisive proportions. The annual income of the American Negro is now more than sixteen billion dollars, almost equal to the national income of Canada. So we are gradually becoming economically independent. It would be a tragic indictment on both the self-respect and practical wisdom of the Negro if history reveals that at the height of the twentieth century the Negro spent more for frivolities than for the cause of freedom. We must never let it be said that we spend more for the evanescent and ephemeral than for the eternal values of freedom and justice.

Another thing that we must do in speeding up the coming of the new age is to develop intelligent, courageous and dedicated leadership. This is one of the pressing needs of the hour. In this period of transition and growing social change, there is a dire need for leaders who are calm and yet positive, leaders who avoid the extremes of "hot-headedness" and "Uncle Tomism." The urgency of the hour calls for leaders of wise judgment and sound integrity—leaders not in love with money, but in love with justice; leaders not in love with publicity, but in love with humanity; leaders who can subject their particular egos to the greatness of the cause. To paraphrase Holland's words:

God give us leaders!
A time like this demands strong minds, great
  hearts,
true faith and ready hands;
Leaders whom the lust of office does not kill;
Leaders whom the spoils of life cannot buy;
Leaders who possess opinions and a will;
Leaders who have honor; leaders who will
  not lie;
Leaders who can stand before a demagogue
and damn his treacherous flatteries without
  winking!
Tall leaders, sun crowned, who live above
  the fog
in public duty and private thinking.

Finally, if we are to speed up the coming of the new age we must have the moral courage to stand up and protest against injustice wherever we find it. Wherever we find segregation we must have the fortitude to passively resist it. I realize that this will mean suffering and sacrifice. It might even mean going to jail. If such is the case we must be willing to fill up the jail houses of the South. It might even mean physical death. But if physical death is the price that some must pay to free their children from a permanent life of psychological death, then nothing could be more honorable. Once more it might well turn out that the blood of the martyr will be the seed of the tabernacle of freedom.

Someone will ask, how will we face the acts of cruelty and violence that might come as results of our standing up for justice? What will be our defense? Certainly it must not be retaliatory violence. We must find our defense in the amazing power of unity and courage that we have demonstrated in Montgomery. Our defense is to meet every act of violence toward an individual Negro with the facts that there are thousands of others who will present themselves in his place as potential victims. Every time one schoolteacher is fired for standing up courageously for justice, it must be faced with the fact that there are four thousand more to be fired. If the oppressors bomb the home of one Negro for his courage, this must be met with the fact that they must be required to bomb the homes of fifty thousand more Negroes. This dynamic unity, this amazing self-respect, this willingness to suffer, and this refusal to hit back will soon cause the oppressor to become ashamed of his own methods. He will be forced to stand before the world and his God splattered with the blood and reeking with the stench of his Negro brother.

There is nothing in all the world greater than freedom. It is worth paying for; it is worth losing a job; it is worth going to jail for. I would rather be a free pauper than a rich slave. I would rather die in abject poverty with my convictions than live in inordinate riches with the lack of self-respect. Once more every Negro must be able to cry out with his forefathers: "Before I'll be a slave, I'll be buried in my grave and go home to my Father and be saved."

If we will join together in doing all of these things we will be able to speed up the coming of the new world—a new world in which men will live together as brothers; a world in which men will beat their swords into ploughshares and their spears into pruning hooks; a world in which men will no longer take necessities from the masses to give luxuries to the classes; a world in which all men will respect the dignity and worth of all human personality. Then we will be able to sing from the great tradition of our nation:

> My Country 'tis of thee,
> Sweet land of liberty,
> Of thee I sing;
> Land where my fathers died;
> Land of the pilgrim's pride;
> From every mountain side
> Let Freedom ring!

This must become literally true. Freedom must ring from every mountainside. Yes, let it ring from the snow-capped Rockies of Colorado, from the prodigious hilltops of New Hampshire, from the mighty Alleghenies of Pennsylvania, from the curvaceous slopes of California. But not only that. Let freedom ring from every mountainside—from every molehill in Mississippi, from Stone Mountain of

Georgia, from Lookout Mountain of Tennessee, yes, and from every hill and mountain of Alabama. From every mountainside let freedom ring. When this day finally comes, "The morning stars will sing together and the sons of God will shout for joy."

Phylon 28 (April 1957): 24–34.

3

The Power of Nonviolence

(1958)

Dr. King offered the following address in response to an invitation from the Young Men's Christian Association (YMCA) and the Young Women's Christian Association (YWCA) at the University of California at Berkeley. Although he spoke on June 4, 1957, before a packed audience of eager students, this excerpt from that address was not published until nearly a year later. But its appearance was timely. Once again, Dr. King asserted his belief in nonviolent direct action despite some ambivalence that some of his followers had about its effectiveness. Would it prove to be strong enough to withstand the violence of white supremacists? Or would events require President Eisenhower to intervene as he did in Little Rock?

Some breathed a sigh of relief when the president removed "federalized" National Guard troops from Little Rock's Central High School on May 8, 1958. The African American community was even more hopeful about the positive consequences of nonviolence when Ernest Green, one of the nine students escorted into Central High School the previous year, graduated from Central on May 27, the only black person in a class of 600. The positive appeal that nonviolent direct action had among some black youth was evident on August 19 when members of the National Association for the Advancement of Colored People (NAACP) Youth Council conducted several sit-ins at lunch counters in Oklahoma City.

One of the most disappointing events of Dr. King's life, and of his effort to promote nonviolence, also occurred in 1958, on September 20. A black woman asked Dr. King "Are you Martin Luther King?" as he was autographing copies of his book *Stride Toward Freedom*. When he answered yes, she stabbed him in the chest. The knife stopped very near his aorta. As he recounted years later in his last public address, "I See the Promised Land," he would have missed the maturation of a great social revolution if he had sneezed while that knife lingered so close to the most vital artery of his heart.

FROM THE VERY BEGINNING THERE WAS A PHILOSOPHY UNDERGIRDING THE Montgomery boycott, the philosophy of nonviolent resistance. There was always the problem of getting this method over because it didn't make sense to most of the people in the beginning. We had to use our mass meetings to explain nonviolence to a community of people who had never heard of the philosophy and in many instances were not sympathetic with it. We had meetings twice a week on Mondays and on Thursdays, and we had an institute on nonviolence and social change. We had to make it clear that nonviolent resistance is not a method of cowardice. It does resist. It is not a method of stagnant passivity and deadening complacency. The nonviolent resister is just as opposed to the evil that he is standing against as the violent resister but he resists without violence. This method is nonaggressive physically but strongly aggressive spiritually.

NOT TO HUMILIATE BUT TO WIN OVER

Another thing that we had to get over was the fact that the nonviolent resister does not seek to humiliate or defeat the opponent but to win his friendship and understanding. This was always a cry that we had to set before people that our aim is not to defeat the white community, not to humiliate the white community, but to win the friendship of all of the persons who had perpetrated this system in the past. The end of violence or the aftermath of violence is bitterness. The aftermath of nonviolence is reconciliation and the creation of a beloved community. A boycott is never an end within itself. It is merely a means to awaken a sense of shame within the oppressor but the end is reconciliation, the end is redemption.

Then we had to make it clear also that the nonviolent resister seeks to attack the evil system rather than individuals who happen to be caught up in the system. And this is why I say from time to time that the struggle in the South is not so much the tension between white people and Negro people. The struggle is rather between justice and injustice, between the forces of light and the forces of darkness. And if there is a victory it will not be a victory merely for fifty thousand Negroes. But it will be a victory for justice, a victory for good will, a victory for democracy.

Another basic thing we had to get over is that nonviolent resistance is also an internal matter. It not only avoids external violence or external physical violence but also internal violence of spirit. And so at the center of our movement stood the philosophy of love. The attitude that the only way to ultimately change humanity and make for the society that we all long for is to keep love at the center of our lives. Now people used to ask me from the beginning what do you mean by love and how is it that you can tell us to love those persons who seek to defeat us and those persons who stand against us; how can you love such persons? And I had to make it clear all along that love in its highest sense is not a sentimental sort of thing, not even an affectionate sort of thing.

AGAPE LOVE

The Greek language uses three words for love. It talks about *eros*. *Eros* is a sort of aesthetic love. It has come to us to be a sort of romantic love and it stands with all of its beauty. But when we speak of loving those who oppose us we're not talking about *eros*. The Greek language talks about *philia* and this is a sort of reciprocal love between personal friends. This is a vital, valuable love. But when we talk of loving those who oppose you and those who seek to defeat you we are not talking about *eros* or *philia*. The Greek language comes out with another word and it is *agape*. *Agape* is understanding, creative, redemptive good will for all men. Biblical theologians would say it is the love of God working in the minds of men. It is an overflowing love which seeks nothing in return. And when you come to love on this level you begin to love men not because they are likable, not because they do things that attract us, but because God loves them and here we love the person who does the evil deed while hating the deed that the person does. It is the type of love that stands at the center of the movement that we are trying to carry on in the Southland—*agape*.

SOME POWER IN THE UNIVERSE THAT WORKS FOR JUSTICE

I am quite aware of the fact that there are persons who believe firmly in nonviolence who do not believe in a personal God, but I think every person who believes in nonviolent resistance believes somehow that the universe in some form is on the side of justice. That there is something unfolding in the universe whether one speaks of it as an unconscious process, or whether one speaks of it as some unmoved mover, or whether someone speaks of it as a personal God. There is something in the universe that unfolds for justice and so in Montgomery we felt somehow that as we struggled we had cosmic companionship. And this was one of the things that kept the people together, the belief that the universe is on the side of justice.

God grant that as men and women all over the world struggle against evil systems they will struggle with love in their hearts, with understanding good will. *Agape* says you must go on with wise restraint and calm reasonableness but you must keep moving. We have a great opportunity in America to build here a great nation, a nation where all men live together as brothers and respect the dignity and worth of all human personality. We must keep moving toward that goal. I know that some people are saying we must slow up. They are writing letters to the North and they are appealing to white people of good will and to the Negroes saying slow up, you're pushing too fast. They are saying we must adopt a policy of moderation. Now if moderation means moving on with wise restraint and calm reasonableness, then moderation is a great virtue that all men of good will must seek to achieve in this tense period of transition. But if moderation means slowing up in the move for justice and capitulating to the whims and caprices of the guardians of the deadening status quo, then moderation is a tragic vice which all men of good will must condemn. We must continue to move on. Our self-respect is at stake; the prestige of our nation is at stake. Civil rights is an eternal moral issue which may well determine the destiny of our civilization in the ideological struggle with communism. We must keep moving with wise restraint and love and with proper discipline and dignity.

THE NEED TO BE "MALADJUSTED"

Modern psychology has a word that is probably used more than any other word. It is the word "maladjusted." Now we all should seek to live a well-adjusted life in order to avoid neurotic and schizophrenic personalities. But there are some things within our social order to which I am proud to be maladjusted and to which I call upon you to be maladjusted. I never intend to adjust myself to segregation and discrimination. I never intend to adjust myself to mob rule. I never intend to adjust myself to the tragic effects of the methods of physical violence and to tragic militarism. I call upon you to be maladjusted to such things. I call upon you to be as maladjusted as Amos who in the midst of the injustices of his day cried out in words that echo across the generation, "Let judgment run down like waters and righteousness like a mighty stream." As maladjusted as Abraham Lincoln who had the vision to see that this nation could not exist half slave and half free. As maladjusted as Jefferson, who in the midst of an age amazingly adjusted to slavery could cry out, "All men are created equal and are endowed by their Creator with certain inalienable rights and that among these are life, liberty and the pursuit of happiness." As maladjusted as Jesus of Nazareth who dreamed a dream of the fatherhood of God and the brotherhood of man. God grant that we will be so maladjusted that we will be able to go out and change our world and our civilization. And then we will be able to move from the bleak and desolate midnight of man's inhumanity to man to the bright and glittering daybreak of freedom and justice.

Intercollegian (May 1958): 8ff.

4

Speech Before the Youth March for Integrated Schools

(1959)

On April 18, 1959, Dr. King, along with several other civil rights leaders, including Daisy Bates, Harry Belafonte, A. Philip Randolph, Jackie Robinson, and Roy Wilkins, spoke before 26,000 high school and college students who had come to Washington, D.C., to demonstrate their support for the 1954 Supreme Court decision against racial segregation in the nation's public schools. This was the second consecutive year that such a march was held. The first march, with 10,000 students present, was held on October 25, 1958.

It was difficult to sustain the idealism of young people about the effectiveness of nonviolence when segregationists seemingly countered every modicum of success with dastardly deeds. For example, Mack Parker, a black youth, was lynched in Popular, Mississippi, on April 25, less than a week after the Youth March for Integrated Schools. As well, such white vigilantism had white collar counterparts. On June 26, the Board of Supervisors of Prince Edward County, Virginia, closed the entire school system rather than allow the county's schools to integrate. This situation deprived black youngsters of an education for many years.

White southerners were not the only ones to use their power to ignore the cry for racial justice. In a public referendum on December 21, white citizens of Deerfield, Illinois, approved a plan that prevented the erection of an interracial housing development.

As I stand here and look out upon the thousands of Negro faces, and the thousands of white faces, intermingled like the waters of a river, I see only one face—the face of the future.

Yes; as I gaze upon this great historic assembly, this unprecedented gathering of young people, I cannot help thinking—that a hundred years from now the historians will be calling this not the "beat" generation, but the generation of integration.

The fact that thousands of you came here to Washington and that thousands more signed your petition proves that this generation will not take "No" for an answer—will not take double talk for an answer—will not take gradualism for an answer. It proves that the only answer you will settle for is—total desegregation and total equality—now.

I know of no words eloquent enough to express the deep meaning, the great power, and the unconquerable spirit back of this inspiringly original, uniquely American march of young people. Nothing like it has ever happened in the history of our nation. Nothing, that is, except the last youth march. What this march demonstrates to me, above all else, is that you young people, through your own experience, have somehow discovered the central fact of American life—that the extension of democracy for all Americans depends upon complete integration of Negro Americans.

By coming here you have shown yourselves to be highly alert, highly responsible young citizens. And very soon the area of your responsibility will increase, for you will begin to exercise your greatest privilege as an American—the right to vote. Of course, you will have no difficulty exercising this privilege—if you are white.

But I wonder if you can understand what it feels like to be a Negro, living in the South, where, by attempting to exercise this right, you may be taking your life in your hands.

The denial of the vote not only deprives the Negro of his constitutional rights—but what is even worse—it degrades him as a human being. And yet, even this degradation, which is only one of many humiliations of everyday life, is losing its ability to degrade. For the southern Negro is learning to transform his degradation into resistance. Nonviolent resistance. And by so doing he is not only achieving his dignity as a human being, he is helping to advance democracy in the South. This is why my colleagues and I in the Southern

Leadership Conference are giving our major attention to the campaign to increase the registration of Negro voters in the South to three million. Do you realize what would happen in this country if we were to gain three million southern Negro votes? We could change the composition of Congress. We could have a Congress far more responsive to the voters' will. We could have all schools integrated—north and south. A new era would open to all Americans. Thus, the Negro, in his struggle to secure his own rights is destined to enlarge democracy for all people, in both a political and a social sense.

Indeed in your great movement to organize a march for integrated schools you have actually accomplished much more. You have awakened on hundreds of campuses throughout the land a new spirit of social inquiry to the benefit of all Americans.

This is really a noble cause. As June approaches, with its graduation ceremonies and speeches, a thought suggests itself. You will hear much about careers, security, and prosperity. I will leave the discussion of such matters to your deans, your principals, and your valedictorians. But I do have a graduation thought to pass along to you. Whatever career you may choose for yourself—doctor, lawyer, teacher—let me propose an avocation to be pursued along with it. Become a dedicated fighter for civil rights. Make it a central part of your life.

It will make you a better doctor, a better lawyer, a better teacher. It will enrich your spirit as nothing else possibly can. It will give you that rare sense of nobility that can only spring from love and selflessly helping your fellow man. Make a career of humanity. Commit yourself to the noble struggle for equal rights. You will make a greater person of yourself, a greater nation of your country, and a finer world to live in.

Congressional Record 105 (May 20, 1959): 8696–97.

PART II

The Dream Enters World History

(1959–1964)

5

My Trip to the Land of Gandhi

(1959)

Martin Luther King, Jr., had a philosophical acquaintance with the philosophy of nonviolence from the time he was an undergraduate at Morehouse College in Atlanta. Dr. Benjamin E. Mays, the president of Morehouse, and the Reverend Dr. Howard Thurman, dean of Chapel and a professor of religion at Morehouse, had both visited Mohandas Gandhi in the late 1930s. Both of these great religious leaders and educators played major roles in encouraging debate about the possibility of using moral means to achieve moral ends. Gandhi's leadership in defying British rule in India through the use of nonviolence made him a significant hero among many progressive Americans on college campuses. Martin Luther King, Jr., was quite aware of Gandhi's achievement. Was it merely coincidence that just a few weeks after Gandhi was assassinated in 1948, King, at the age of 18, was ordained by Ebenezer Baptist Church in Atlanta to the Christian ministry? We may never know the answer to this question. But we do know that during the month of September 1948 two devotees of nonviolent direct action, Dr. Abraham J. Muste, head of the Fellowship of Reconciliation, and the Reverend Dr. Mordecai Johnson, president of Howard University, preached on the life and teachings of Mahatma Gandhi at Crozer Theological Seminary in Chester, Pennsylvania. Martin Luther King, Jr., had just entered that notable Baptist seminary to work toward his

Bachelor of Divinity degree. There King began a serious study of Gandhi. But he was not convinced about how well nonviolence would work in the midst of the violence of white racists in the United States.

During the Montgomery campaign in 1956, the Reverend Glen Smiley and Bayard Rustin, both representatives of the Fellowship of Reconciliation, convinced Dr. King to make nonviolent direct action the philosophical center of the Montgomery movement. Indeed, this philosophy became the intellectual bedrock of the Southern Christian Leadership Conference when it was formed in 1957.

Between February 2 and March 10, 1959, Dr. and Mrs. King toured India with Dr. Lawrence D. Reddick, a black professor of history at Alabama State University in Montgomery, to study Gandhi's philosophy and techniques of nonviolence. They were guests of India's Prime Minister Jawaharlal Nehru, one of Gandhi's disciples.

FOR A LONG TIME I HAD WANTED TO TAKE A TRIP TO INDIA. EVEN AS A child the entire Orient held a strange fascination for me—the elephants, the tigers, the temples, the snake charmers and all the other storybook characters.

While the Montgomery boycott was going on, India's Gandhi was the guiding light of our technique of nonviolent social change. We spoke of him often. So as soon as our victory over bus segregation was won, some of my friends said: "Why don't you go to India and see for yourself what the Mahatma, whom you so admire, has wrought."

In 1956 when Pandit Jawaharlal Nehru, India's prime minister, made a short visit to the United States, he was gracious enough to say that he wished that he and I had met and had his diplomatic representatives make inquiries as to the possibility of my visiting his country some time soon. Our former American ambassador to India, Chester Bowles, wrote me along the same lines.

But every time that I was about to make the trip, something would interfere. At one time it was my visit by prior commitment to Ghana. At another time my publishers were pressing me to finish writing *Stride Toward Freedom*. Then along came Mrs. Izola Ware Curry. When she struck me with that Japanese letter opener on that

Saturday afternoon in September as I sat autographing books in a Harlem store, she not only knocked out the travel plans that I had but almost everything else as well.

After I recovered from this near-fatal encounter and was finally released by my doctors, it occurred to me that it might be better to get in the trip to India before plunging too deeply once again into the sea of the southern segregation struggle.

I preferred not to take this long trip alone and asked my wife and my friend, Lawrence Reddick, to accompany me. Coretta was particularly interested in the women of India and Dr. Reddick in the history and government of that great country. He had written my biography, *Crusader Without Violence,* and said that my true test would come when the people who knew Gandhi looked me over and passed judgment upon me and the Montgomery movement. The three of us made up a sort of three-headed team with six eyes and six ears for looking and listening.

The Christopher Reynolds Foundation made a grant through the American Friends Service Committee to cover most of the expenses of the trip and the Southern Christian Leadership Conference and the Montgomery Improvement Association added their support. The Gandhi Memorial Trust of India extended an official invitation, through diplomatic channels, for our visit.

And so on February 3, 1959, just before midnight, we left New York by plane. En route we stopped in Paris with Richard Wright, an old friend of Reddick's, who brought us up to date on European attitudes on the Negro question and gave us a taste of the best French cooking.

We missed our plane connection in Switzerland because of fog, arriving in India after a roundabout route, two days late. But from the time we came down out of the clouds at Bombay on February 10, until March 10, when we waved goodbye at the New Delhi airport, we had one of the most concentrated and eye-opening experiences of our lives. There is so much to tell that I can only touch upon a few of the high points.

At the outset, let me say that we had a grand reception in India. The people showered upon us the most generous hospitality imaginable. We were graciously received by the prime minister, the president and the vice-president of the nation; members of Parliament, governors and chief ministers of various Indian states; writers, professors, social reformers and at least one saint. Since our pictures were in the newspapers very often it was not unusual for us to be recognized by crowds in public places and on public conveyances. Occasionally I would take a morning walk in the large cities, and out of the most unexpected places someone would emerge and ask: "Are you Martin Luther King?"

Virtually every door was open to us. We had hundreds of invitations that the limited time did not allow us to accept. We were looked upon as brothers with the color of our skins as something of an asset. But the strongest bond of fraternity was the common cause of minority and colonial peoples in America, Africa and Asia struggling to throw off racialism and imperialism.

We had the opportunity to share our views with thousands of Indian people through endless conversations and numerous discussion sessions. I spoke before university groups and public meetings all over India. Because of the keen interest that the Indian people have in the race problem these meetings were usually packed. Occasionally interpreters were used, but on the whole I spoke to audiences that understood English.

The Indian people love to listen to the Negro spirituals. Therefore, Coretta ended up singing as much as I lectured. We discovered that autograph seekers are not confined to America. After appearances in public meetings and while visiting villages we were often besieged for autographs. Even while riding planes, more than once pilots came into the cabin from the cockpit requesting our signatures.

We got a good press throughout our stay. Thanks to the Indian papers, the Montgomery bus boycott was already well known in that country. Indian publications perhaps gave a better continuity of our 381-day bus strike than did most of our papers in the United States. Occasionally I meet some American fellow citizen who even now asks me how the bus boycott is going, apparently never having read that our great day of bus integration, December 21, 1956, closed that chapter of our history.

We held press conferences in all of the larger cities—Delhi, Calcutta, Madras and Bombay—and talked with newspapermen almost everywhere we went. They asked sharp questions and at times appeared to be hostile but that was just their way of bringing out the story that they were after. As reporters, they were scrupulously fair with us and in their editorials showed an amazing grasp of what was going on in America and other parts of the world.

The trip had a great impact upon me personally. It was wonderful to be in Gandhi's land, to talk with his son, his grandsons, his cousins and other relatives; to share the reminiscences of his close comrades, to visit his ashrama, to see the countless memorials for him and finally to lay a wreath on his entombed ashes at Rajghat. I left India more convinced than ever before that nonviolent resistance is the most potent weapon available to oppressed people in their struggle for freedom. It was a marvelous thing to see the amazing results of a nonviolent campaign. The aftermath of hatred and bitterness that usually follows a violent campaign was found nowhere in India. Today a mutual friendship based on complete equality exists between the Indian and British people within the commonwealth. The way of acquiescence leads to moral and spiritual suicide. The way of violence leads to bitterness in the survivors and brutality in the destroyers. But, the way of nonviolence leads to redemption and the creation of the beloved community.

The spirit of Gandhi is very much alive in India today. Some of his disciples have misgivings about this when they remember the drama of the fight for national independence and when they look around and find nobody today who comes near the stature of the Mahatma. But any objective observer must report that Gandhi is not only the greatest figure in India's history but that his influence is felt in almost every aspect of life and public policy today.

India can never forget Gandhi. For example, the Gandhi Memorial Trust (also known as the Gandhi Smarak Nidhi) collected some $130 million soon after the death of "the father of the nation." This was perhaps the largest, spontaneous, mass monetary contribution to the memory of a single individual in the history of the world. This fund, along with support from the Government and other institutions, is resulting in the spread and development of Gandhian philosophy, the implementing of his constructive program, the erection of libraries and the publication of works by and about the life and times of Gandhi. Posterity could not escape him even if it tried. By all standards of measurement, he is one of the half-dozen greatest men in world history.

I was delighted that the Gandhians accepted us with open arms. They praised our experiment with the nonviolent resistance technique at Montgomery. They seem to look upon it as an outstanding example of the possibilities of its use in Western civilization. To them as to me it also suggests that nonviolent resistance *when planned and positive in action* can work effectively even under totalitarian regimes.

We argued this point at some length with the groups of African students who are today studying in India. They felt that nonviolent resistance could only work in a situation where the resisters had a potential ally in the conscience of the opponent. We soon discovered that they, like many others, tended to confuse passive resistance with nonresistance. This is completely wrong. True nonviolent resistance is not unrealistic submission to evil power. It is rather a courageous confrontation of evil by the power of love, in the faith that it is better to be the recipient of violence than the inflictor of it, since the latter only multiplies the existence of violence and bitterness in the universe, while the former may develop a sense of shame in the opponent, and thereby bring about a transformation and change of heart.

Nonviolent resistance does call for love, but it is not a sentimental love. It is a very stern love that would organize itself into collective action to right a wrong by taking on itself suffering. While I understand the reasons why oppressed people often turn to violence in their struggle for freedom, it is my firm belief that the crusade for independence and human dignity that is now reaching a climax in Africa will have a more positive effect on the world, if it is waged along the lines that were first demonstrated in that continent by Gandhi himself.

India is a vast country with vast problems. We flew over the long stretches, from north to south, east to west; took trains for shorter jumps and used automobiles and jeeps to get us into the less accessible places.

India is about a third the size of the United States but has almost three times as many people. Everywhere we went we saw crowded humanity—on the roads, in the city streets and squares, even in the villages.

Most of the people are poor and poorly dressed. The average income per person is less than seventy dollars per year. Nevertheless, their turbans for their heads, loose-flowing, wrap-around *dhotis* that they wear instead of trousers and the flowing saris that the women wear instead of dresses are colorful and picturesque. Many Indians wear part native and part Western dress.

We think that we in the United States have a big housing problem, but in the city of Bombay, for example, over a half-million people sleep out of doors every night. These are mostly unattached, unemployed or partially employed males. They carry their bedding with them like foot soldiers and unroll it each night in any unoccupied space they can find—on the sidewalk, in a railroad station or at the entrance of a shop that is closed for the evening.

The food shortage is so widespread that it is estimated that less than thirty percent of the people get what we would call three square meals a day. During our great depression of the 1930s, we spoke of "a third of a nation" being "ill-housed, ill clad and ill fed." For India today, simply change one-third to two-thirds in that statement and that would make it about right.

As great as is unemployment, under-employment is even greater. Seventy percent of the Indian people are classified as agricultural workers and most of these do less than two hundred days of farm labor per year because of the seasonal fluctuations and other uncertainties of mother nature. Jobless men roam the city streets.

Great ills flow from the poverty of India but strangely there is relatively little crime. Here is another concrete manifestation of the wonderful spiritual quality of the Indian people. They are poor, jammed together and half starved but they do not take it out on each other. They are a kindly people. They do not abuse each other—verbally or physically—as readily as we do. We saw but one fist fight in India during our stay.

In contrast to the poverty-stricken, there are Indians who are rich, have luxurious homes, landed estates, fine clothes and show evidence of overeating. The bourgeoisie—white, black or brown—behaves about the same the world over.

And then there is, even here, the problem of segregation. We call it race in America; they call it caste in India. In both places it means that some are considered inferior, treated as though they deserve less.

We were surprised and delighted to see that India has made greater progress in the fight against caste "untouchability" than we have made here in our own country against race segregation. Both nations have federal laws against discrimination (acknowledging, of course, that the decision of our Supreme Court is the law of our land). But after this has been said, we must recognize that there are great differences between what India has done and what we have done on a problem that is very similar. The leaders of India have placed their moral power behind their law. From the Prime Minister down to the village councilmen, everybody declares publicly that untouchability is wrong. But in the United States some of our highest officials decline to render a moral judgment on segregation and some from the South publicly boast of their determination to maintain segregation. This would be unthinkable in India.

Moreover, Gandhi not only spoke against the caste system but he acted against it. He took "untouchables" by the hand and led them into the temples from which they had been excluded. To equal that, President Eisenhower would take a Negro child by the hand and lead her into Central High School in Little Rock.

Gandhi also renamed the untouchables, calling them "Harijans" which means "children of God."

The government has thrown its full weight behind the program of giving the Harijans an equal chance in society—especially when it comes to job opportunities, education, and housing.

India's leaders, in and out of government, are conscious of their country's other great problems and are heroically grappling with them. The country seems to be divided. Some say that India should become Westernized and modernized as quickly as possible so that she might raise her standards of living. Foreign capital and foreign industry should be invited in, for in this lies the salvation of the almost desperate situation.

On the other hand, there are others—perhaps the majority—who say that Westernization will bring with it the evils of materialism, cutthroat competition and rugged individualism; that India will lose her soul if she takes to chasing Yankee dollars; and that the big machine will only raise the living standards of the comparative few workers who get jobs but that the greater number of people will be displaced and will thus be worse off than they are now.

Prime Minister Nehru, who is at once an intellectual and a man charged with the practical responsibility of heading the government, seems to steer a middle course between these extreme attitudes. In our talk with him he indicated that he felt that some industrialization was absolutely necessary; that there were some things that only big or heavy industry could do for the country but that if the state keeps a watchful eye on the developments, most of the pitfalls may be avoided.

At the same time, Mr. Nehru gives support to the movement that would encourage and expand the handicraft arts such as spinning and weaving in home and village and thus leave as much economic self-help and autonomy as possible to the local community.

There is a great movement in India that is almost unknown in America. At its center is the campaign for land reform known as Bhoodan. It would solve India's great economic and social change by consent, not by force. The Bhoodanists are led by the saints Vinoba Bhave and Jayaprakash Narayan, a highly sensitive intellectual, who was trained in American colleges. Their ideal is self-sufficiency. Their program envisions:

1. *Persuading* large landowners to give up some of their holding to landless peasants;
2. *Persuading* small landowners to give up their individual ownership for common cooperative ownership by the villages;
3. *Encouraging* farmers and villagers to spin and weave the cloth for their own clothes during their spare time from their agricultural pursuits.

Since these measures would answer the questions of employment, food and clothing, the village could then, through cooperative action, make just about everything that it would need or get it through bartering and exchange from other villages. Accordingly, each village would be virtually self-sufficient and would thus free itself from the domination of the urban centers that are today like evil lodestones drawing the people away from the rural areas, concentrating them in city slums and debauching them with urban vices. At least this is the argument of the Bhoodanists and other Gandhians.

Such ideas sound strange and archaic to Western ears. However, the Indians have already achieved greater results than we Americans would ever expect. For example, millions of acres of land have been given up by rich landlords and additional millions of acres have been given up to cooperative management by small farmers. On the other hand, the Bhoodanists shrink from giving their movement the organization and drive that we in America would venture to guess that it must have in order to keep pace with the magnitude of the problems that everybody is trying to solve.

Even the government's five-year plans fall short in that they do not appear to be of sufficient scope to embrace their objectives. Thus, the three five-year plans were designed to provide twenty-five million new jobs over a fifteen-year period but the birth rate of India is six million per year. This means that in fifteen years there will be nine million more people (less those who have died or retired) looking for the fifteen million new jobs. In other words, if the planning were one hundred percent successful, it could not keep pace with the growth of problems it is trying to solve.

As for what should be done, we surely do not have the answer. But we do feel certain that India needs help. She must have outside capital and technical know-how. It is in the interest of the United States and the West to help supply these needs and *not attach strings to the gifts.*

Whatever we do should be done in a spirit of international brotherhood, not national selfishness. It should be done not merely because it is diplomatically expedient, but because it is morally correct. At the same time, it will rebound to the credit of the West if India is able to maintain her democracy while solving her problems.

It would be a boon to democracy if one of the great nations of the world, with almost four hundred million people, proves that it is possible to provide a good living for everyone without surrendering to a dictatorship of either the "right" or "left." Today India is a tremendous force for peace and nonviolence, at home and abroad. It is a land where the idealist and the intellectual are yet respected. We should want to help India preserve her soul and thus help to save our own.

Ebony (July 1959): 84–86, 88–90, 92. This article was written with the editorial assistance of Reddick, who also did the primary research for King's *Stride Toward Freedom*.

6

The Social Organization of Nonviolence

(1959)

While Dr. King's trip to India underscored his commitment to nonviolence, he still struggled to keep pace with the rapidly increasing disenchantment within the black community, especially among young people. He was caught between racial moderates who moved too slowly, and white segregationists and black militants who were prepared to fight a racial civil war. In the following essay, Dr. King argued against Robert Williams, a die-hard advocate of militant black self-defense. Williams, the local president of the National Association for the Advancement of Colored People (NAACP) in Monroe County, North Carolina, and an ex-marine, formed, drilled, and armed blacks to defend themselves. Both the NAACP and Southern Christian Leadership Conference (SCLC) strongly denounced Williams's endorsement of violence. But events were beyond anyone's control.

The nation was also awaiting the results of the presidential campaign between the Republican Richard Nixon and the Democrat John Kennedy that would conclude in November 1960. In the meantime, on November 29, 1959, Dr. King submitted his resignation as pastor of the Dexter Avenue Baptist Church in Montgomery, Alabama, in order to devote himself full-time to the presidency of the SCLC.

PARADOXICALLY, THE STRUGGLE FOR CIVIL RIGHTS HAS REACHED A STAGE of profound crisis, although its outward aspect is distinctly less turbulent and victories of token integration have been won in the hard-resistance areas of Virginia and Arkansas.

The crisis has its origin in a decision rendered by the Supreme Court more than a year ago which upheld the pupil placement law. Though little noticed then, this decision fundamentally weakened the historic 1954 ruling of the Court. It is imperceptibly becoming the basis of a *de facto* compromise between the powerful contending forces.

The 1954 decision required for effective implementation resolute federal action supported by mass action to undergird all necessary changes. It is obvious that federal action by the legislative and executive branches was half-hearted and inadequate. The activity of Negro forces, while heroic in some instances, and impressive in other sporadic situations, lacked consistency and militancy sufficient to fill the void left by government default. The segregationists were swift to seize these advantages, and unrestrained by moral or social conscience, defied the law boldly and brazenly.

The net effect of this social equation has led to the present situation, which is without clearcut victory for either side. Token integration is a developing pattern. This type of integration is merely an affirmation of a principle without the substance of change.

It is, like the Supreme Court decision, a pronouncement of justice, but by itself does not insure that the millions of Negro children will be educated in conditions of equality. This is not to say that it is without value. It has substantial importance. However, it fundamentally changes the outlook of the whole movement, for it raises the prospect of long, slow change without a predictable end. As we have seen in northern cities, token integration has become a pattern in many communities and remained frozen, even though environmental attitudes are substantially less hostile to full integration than in the South.

THREE VIEWS OF VIOLENCE

This then is the danger. Full integration can easily become a distant or mythical goal—major integration may be long postponed, and in the quest for social calm a compromise firmly implanted in which the real goals are merely token integration for a long period to come.

The Negro was the tragic victim of another compromise in 1878, when his full equality was bargained away by the federal government and a condition somewhat above slave status but short of genuine citizenship became his social and political existence for nearly a century.

There is reason to believe that the Negro of 1959 will not accept supinely any such compromises in the contemporary struggle for integration. His struggle will continue, but the obstacles will determine its specific nature. It is axiomatic in social life that the imposition of frustrations leads to two kinds of reactions. One is the development of a wholesome social organization to resist with effective, firm measures any efforts to impede progress. The other is a confused, anger-motivated drive to strike back violently, to inflict damage. Primarily, it seeks to cause injury to retaliate for wrongful suffering. Secondarily, it seeks real progress. It is punitive—not radical or constructive.

The current calls for violence have their roots in this latter tendency. Here one must be clear that there are three different views on the subject of violence. One is the approach of pure nonviolence, which cannot readily or easily attract large masses, for it requires extraordinary discipline and courage. The second is violence exercised in self-defense, which all societies, from the most primitive to the most cultured and civilized, accept as moral and legal. The principle of self-defense, even involving weapons and bloodshed, has never been condemned, even by Gandhi, who sanctioned it for those unable to master pure nonviolence. The third is the advocacy of violence as a tool of advancement, organized as in warfare, deliberately and consciously. To this tendency many Negroes are being tempted today. There are incalculable perils in this approach. It is not the danger or sacrifice of physical being which is primary, though it cannot be contemplated without a sense of deep concern for human life. The greatest danger is that it will fail to attract Negroes to a real collective struggle, and will confuse the large uncommitted middle group, which as yet has not supported either side. Further, it will mislead Negroes into the belief that this is the only path and place them as a minority in a position where they confront a far larger adversary than it is possible to defeat in this form of combat. When the Negro uses force in self-defense he does not forfeit support—he may even win it, by the courage and self-respect it reflects. When he seeks to initiate violence he provokes questions about the necessity for it, and inevitably is blamed for its consequences. It is unfortunately true that however the Negro acts, his struggle will not be free of violence initiated by his enemies, and he will need ample courage and willingness to sacrifice to defeat this manifestation of violence. But if he seeks it and organizes it, he cannot win. Does this leave the Negro without a positive method to advance? Mr. Robert Williams would have us believe that there is no effective and practical alternative. He argues that we must be cringing and submissive or take up arms. To so place the issue distorts the whole problem. There are other meaningful alternatives.

The Negro people can organize socially to initiate many forms of struggle which can drive their enemies back without resort to futile and harmful violence. In the history of the movement for racial advancement, many creative forms have been developed—the mass boycott, sit-down protests and strikes, sit-ins—refusal to pay fines and bail for unjust arrests—mass marches—mass meetings—prayer pilgrimages, etc. Indeed, in Mr. Williams' own community of Monroe, North Carolina, a striking example of collective community action won a significant victory without use of arms or threats of violence. When the police incarcerated a Negro doctor unjustly, the aroused people of Monroe marched to the police station, crowded into its halls and corridors, and refused to leave until their colleague was released. Unable to arrest everyone, the authorities released the doctor and neither side attempted to unleash violence. This experience was related by the doctor who was the intended victim.

There is more power in socially organized masses on the march than there is in guns in the hands of a few desperate men. Our enemies would prefer to deal with a small armed group rather than with a huge, unarmed but resolute mass of people. However, it is necessary that the mass-action method be persistent and unyielding. Gandhi said the Indian people must "never let them rest," referring to the British. He urged them to keep protesting daily and weekly, in a variety of ways. This method inspired and organized the Indian masses and disorganized and demobilized the British. It educates its myriad participants, socially and morally. All history teaches us that like a turbulent ocean beating great cliffs into fragments of rock, the determined movement of people incessantly demanding their rights always disintegrates the old order.

It is this form of struggle—non-cooperation with evil through mass actions—"never letting them rest"—which offers the more effective road for those who have been tempted and goaded to violence. It needs the bold and the brave because it is not free of danger. It faces the vicious and evil enemies squarely. It requires dedicated people, because it is a backbreaking task to arouse, to organize, and to educate tens of thousands for disciplined, sustained action. From this form of struggle more emerges that is permanent and damaging to the enemy than from a few acts of organized violence.

Our present urgent necessity is to cease our internal fighting and turn outward to the enemy—using every form of mass action yet known—create new forms—and resolve never to let them rest. This is the social lever which will force open the door to freedom. Our powerful weapons are the voices, the feet, and the bodies of dedicated, united people, moving without rest toward a just goal. Greater tyrants than southern segregationists have been subdued and defeated by this form of struggle. We have not yet used it, and it would be tragic if we spurn it because we have failed to perceive its dynamic strength and power.

CASHING IN ON WAR?

I am reluctant to inject a personal defense against charges by Mr. Williams that I am inconsistent in my struggle against war and too weak-kneed to protest nuclear war. Merely to set the record straight, may I state that repeatedly, in public addresses and in my writings, I have unequivocally declared my hatred for this most colossal of all evils and I have condemned any organizer of war, regardless of his rank or nationality. I have signed numerous statements with other Americans condemning nuclear testing and have authorized publication of my name in advertisements appearing in the largest-circulation newspapers in the country, without concern that it was then "unpopular" to so speak out.

Liberation (October 1959): 5–6.

7

Pilgrimage to Nonviolence

(1960)

Dr. King celebrated his thirty-first birthday on January 15, 1960. A few days later he and his family moved to Atlanta, Georgia, to expand his duties as president of the Southern Christian Leadership Conference. He also became the co-pastor of the Ebenezer Baptist Church in Atlanta.

Dr. and Mrs. King had barely moved their belongings into their new home before the Freedom Movement entered a new and more radical phase called the "Sit-In Movement." "Sit-ins" were attempts to attack laws and customs that prohibited African American people from eating at lunch counters or restaurants alongside Caucasians. In this form of nonviolent resistance, groups of black patrons simply took seats wherever they were prohibited from doing so because of their race. Primarily African American college students led this phase of the Civil Rights Movement. On February 1, four black students at North Carolina A & T College in Greensboro, North Carolina, started a sit-in in a local "five and dime" store. By February 10, this movement spread among college students in fifteen cities in at least five southern states. A race riot occurred during one of these sit-in demonstrations in Chattanooga, Tennessee, on February 23. One hundred students were arrested in Nashville, Tennessee, on February 27.

At least a thousand students marched toward the Alabama State Capitol Building in Montgomery on March 1. The next day Alabama's Board of Education expelled nine of the student demonstrators. These "educators" did not encourage intellectual debate. The intellectual argument for both integration and nonviolence had not penetrated much of American higher education. But such discussions were widely current within theological circles.

The following article appeared in the weekly magazine *Christian Century* as part of its famous series on "How My Mind Has Changed," wherein world-renowned theologians were given the opportunity to reflect on or update their theological views. Here Dr. King discusses how he came to the intellectual decision to embrace nonviolence.

TEN YEARS AGO I WAS JUST ENTERING MY SENIOR YEAR IN THEOLOGICAL seminary. Like most theological students I was engaged in the exciting job of studying various theological theories. Having been raised in a rather strict fundamentalistic tradition, I was occasionally shocked as my intellectual journey carried me through new and sometimes complex doctrinal lands. But despite the shock the pilgrimage was always stimulating, and it gave me a new appreciation for objective appraisal and critical analysis. My early theological training did the same for me as the reading of Hume did for Kant: it knocked me out of my dogmatic slumber.

At this stage of my development I was a thoroughgoing liberal. Liberalism provided me with an intellectual satisfaction that I could never find in fundamentalism. I became so enamored of the insights of liberalism that I almost fell into the trap of accepting uncritically everything that came under its name. I was absolutely convinced of the natural goodness of man and the natural power of human reason.

I

The basic change in my thinking came when I began to question some of the theories that had been associated with so-called liberal theology. Of course there is one phase of liberalism that I hope to cherish always: its devotion to the search for truth, its insistence on an open and analytical mind, its refusal to abandon the best light of reason. Liberalism's contribution to the philological-historical criticism of biblical literature has been of immeasurable value and should be defended with religious and scientific passion.

It was mainly the liberal doctrine of man that I began to question. The more I observed the tragedies of history and man's shameful inclination to choose the low road, the more I came to see the depths and strength of sin. My reading of the works of Reinhold Niebuhr made me aware of the complexity of human motives and the reality of sin on every level of man's existence. Moreover, I came to recognize the complexity of man's social involvement and the glaring reality of collective evil. I came to feel that liberalism had been all too sentimental concerning human nature and that it leaned toward a false idealism.

I also came to see that liberalism's superficial optimism concerning human nature caused it to overlook the fact that reason is darkened by sin. The more I thought about human nature the more I saw how our tragic inclination for sin causes us to use our minds to rationalize our actions. Liberalism failed to see that reason by itself is little more than an instrument to justify man's defensive ways of thinking. Reason, devoid of purifying power of faith, can never free itself from distortions and rationalizations.

In spite of the fact that I had to reject some aspects of liberalism, I never came to an all-out acceptance of neo-orthodoxy. While I saw neo-orthodoxy as a helpful corrective for a liberalism that had become all too sentimental, I never felt that it provided an adequate answer to the basic questions. If liberalism was too optimistic concerning human nature, neo-orthodoxy was too pessimistic. Not only on the question of man but also on other vital issues neo-orthodoxy went too far in its revolt. In its attempt to preserve the transcendence of God, which had been neglected by liberalism's overstress of his immanence, neo-orthodoxy went to the extreme of stressing a God who was hidden, unknown and "wholly other." In its revolt against liberalism's overemphasis on the power of reason, neo-orthodoxy fell into a mood of antirationalism and semifundamentalism, stressing a narrow, uncritical biblicism. This approach, I felt, was inadequate both for the church and for personal life.

So although liberalism left me unsatisfied on the question of the nature of man, I found no refuge in neo-orthodoxy. I am now convinced that the truth about man is found neither in liberalism nor in neo-orthodoxy. Each represents a partial truth. A large segment of Protestant liberalism defined man only in terms of his essential nature, his capacity for good. Neo-orthodoxy tended to define man only in terms of his existential nature, his capacity for evil. An adequate understanding of man is found neither in the thesis of liberalism nor in the antithesis of neo-orthodoxy, but in a synthesis which reconciles the truths of both.

During the past decade I also gained a new appreciation for the philosophy of existentialism. My first contact with this philosophy came through my reading of Kierkegaard and Nietzsche. Later I turned to a study of Jaspers, Heidegger and Sartre. All of these thinkers stimulated my thinking; while finding things to question in each, I nevertheless learned a great deal from study of them. When I finally turned to a serious study of the works of Paul Tillich I became convinced that existentialism, in spite of the fact that it had become all too fashionable, had grasped certain basic truths about man and his condition that could not be permanently overlooked.

Its understanding of the "finite freedom" of man is one of existentialism's most lasting contributions, and its perception of the anxiety and conflict produced in man's personal and social life as a result of the perilous and ambiguous structure of existence is especially meaningful for our time. The common point in all existentialism, whether it is atheistic or theistic, is that man's existential situation is a state of estrangement from his essential nature. In their revolt against Hegel's essentialism, all existentialists contend that the world is fragmented. History is a series of unreconciled conflicts and man's existence is filled with anxiety and threatened with meaninglessness. While the ultimate Christian answer is not found in any of these existential assertions, there is much here that the theologian can use to describe the true state of man's existence.

Although most of my formal study during this decade has been in systematic theology and philosophy, I have become more and more interested in social ethics. Of course my concern for social problems was already substantial before the beginning of this decade. From my early teens in Atlanta I was deeply concerned about the problem of racial injustice. I grew up abhorring segregation, considering it both rationally inexplicable and morally unjustifiable. I could never accept the fact of having to go to the back of a bus or sit in the segregated section of a train. The first time that I was seated behind a curtain in a dining car I felt as if the curtain had been dropped on my selfhood. I had also learned that the inseparable twin of racial injustice is economic injustice. I saw how the systems of segregation ended up in the exploitation of the Negro as well as the poor whites. Through these early experiences I grew up deeply conscious of the varieties of injustice in our society.

II

Not until I entered theological seminary, however, did I begin a serious intellectual quest for a method to eliminate social evil. I was immediately influenced by the social gospel. In the early fifties I read Rauschenbusch's *Christianity and the Social Crisis,* a book which left an indelible imprint on my thinking. Of course there were points at which I differed with Rauschenbusch. I felt that he had fallen victim to the nineteenth century "cult of inevitable progress," which led him to an unwarranted optimism concerning human nature. Moreover, he came perilously close to identifying the kingdom of God with a particular social and economic system—a temptation which the church should never give in to. But in spite of these shortcomings Rauschenbusch gave to American Protestantism a sense of social responsibility that it should never lose. The gospel at its best deals with the whole man, not only his soul but his body, not only his spiritual well-being, but his material well-being. Any religion that professes to be concerned about the souls of men and is not concerned about the slums that damn them, the economic conditions that strangle them and the social conditions that cripple them is a spiritually moribund religion awaiting burial.

After reading Rauschenbusch I turned to a serious study of the social and ethical theories of the great philosophers. During this period I had almost despaired of the power of love in solving social problems. The "turn the other cheek" philosophy and the "love your enemies" philosophy are only valid, I felt, when individuals are in conflict with other individuals; when racial groups and nations are in conflict a more realistic approach is necessary. Then I came upon the life and teachings of Mahatma Gandhi. As I read his works I became deeply fascinated by his campaigns of nonviolent resistance. The whole Gandhian concept of *satyagraha* (*satya* is truth which equals love, and *graha* is force; *satyagraha* thus means truth-force or love-force) was profoundly significant to me. As I delved deeper into the philosophy of Gandhi my skepticism concerning the power of love gradually diminished, and I came to see for the first time that the Christian doctrine of love operating through the Gandhian method of nonviolence was one of the most potent weapons available to oppressed people in their struggle for freedom. At this time, however, I had a merely intellectual understanding and appreciation of the position, with no firm determination to organize it in a socially effective situation.

When I went to Montgomery, Alabama, as a pastor in 1954, I had not the slightest idea that I would later become involved in a crisis in which nonviolent resistance would be applicable. After I had lived in the community about a year, the bus boycott began. The Negro people of Montgomery, exhausted by the humiliating experiences that they had constantly faced on the buses, expressed in a massive act of noncooperation their determination to be free. They came to see that it was ultimately more honorable to walk the streets in dignity than to ride the buses in humiliation. At the beginning of the protest the people called on me to serve as their spokesman. In accepting this responsibility my mind, consciously or unconsciously, was driven back to the Sermon on the Mount and the Gandhian method of nonviolent resistance. This principle became the guiding light of our movement. Christ furnished the spirit and motivation while Gandhi furnished the method.

The experience in Montgomery did more to clarify my thinking on the question of nonviolence than all of the books that I had read. As the days unfolded I became more and more convinced of the power of nonviolence. Living through the actual experience of the protest, nonviolence became more than a method to which I gave intellectual assent; it became a commitment to a way of life. Many issues I had not cleared up intellectually concerning nonviolence were now solved in the sphere of practical action.

A few months ago I had the privilege of traveling to India. The trip had a great impact on me personally and left me even more convinced of the power of nonviolence. It was a marvelous thing to see the amazing results of a nonviolent struggle. India won her independence, but without violence on the part of Indians. The aftermath of hatred and bitterness that usually follows a violent campaign is found nowhere in India. Today a mutual friendship based on complete equality exists between the Indian and British people within the commonwealth.

I do not want to give the impression that nonviolence will work miracles overnight. Men are not easily moved from their mental ruts or purged of their prejudiced and irrational feelings. When the underprivileged demand freedom, the privileged first react with bitterness and resistance. Even when the demands are couched in nonviolent terms, the initial response is the same. I am sure that many of our white brothers in Montgomery and across the South are still bitter toward Negro leaders, even though these leaders have sought to follow a way of love and nonviolence. So the nonviolent approach does not immediately change the heart of the oppressor. It first does something to the hearts and souls of those committed to it. It gives them new self-respect; it calls up resources of strength and courage that they did not know they had. Finally, it reaches the opponent and so stirs his conscience that reconciliation becomes a reality.

III

During recent months I have come to see more and more the need for the method of nonviolence in international relations. While I was convinced during my student days of the power of nonviolence in group conflicts within nations, I was not yet convinced of its efficacy in conflicts between nations. I felt that while war could never be a positive or absolute good, it could serve as a negative good in the sense of preventing the spread and growth of an evil force. War, I felt, horrible as it is, might be preferable to surrender to a totalitarian system. But more and more I have come to the conclusion that the potential destructiveness of modern weapons of war totally rules out the possibility of war ever serving again as a negative good. If we assume that mankind has a right to survive then we must find an alternative to war and destruction. In a day when sputniks dash through outer space and guided ballistic missiles are carving highways of death through the stratosphere, nobody can win a war. The choice today is no longer between violence and nonviolence. It is either nonviolence or nonexistence.

I am no doctrinaire pacifist. I have tried to embrace a realistic pacifism. Moreover, I see the pacifist position not as sinless but as the lesser evil in the circumstances. Therefore I do not claim to be free from the moral dilemmas that the Christian nonpacifist confronts. But I am convinced that the church cannot remain silent while mankind faces the threat of being plunged into the abyss of nuclear annihilation. If the church is true to its mission it must call for an end to the arms race.

In recent months I have also become more and more convinced of the reality of a personal God. True, I have always believed in the personality of God. But in past years the idea of a personal God was little more than a metaphysical category which I found theologically and philosophically satisfying. Now it is a living reality that has been validated in the experiences of everyday life. Perhaps the suffering, frustration and agonizing moments which I have had to undergo occasionally as a result of my involvement in a difficult struggle have drawn me closer to God. Whatever the cause, God has been profoundly real to me in recent months. In the midst of outer dangers I have felt an inner calm and known resources of strength that only God could give. In many instances I have felt the power of God transforming the fatigue of despair into the buoyancy of hope. I am convinced that the universe is under the control of a loving purpose and that in the struggle for righteousness man has cosmic companionship. Behind the harsh appearances of the world there is a benign power. To say God is personal is not to make him an object among other objects or attribute to him the finiteness and limitations of human personality; it is to take what is finest and noblest in our consciousness and affirm its perfect existence in him. It is certainly true that human personality is limited, but personality as such involves no necessary limitations. It simply means self-consciousness and self-direction. So in the truest sense of the word, God is a living God. In him there is feeling and will, responsive to the deepest yearnings of the human heart: thus God both evokes and answers prayers.

The past decade has been a most exciting one. In spite of the tensions and uncertainties of our age something profoundly meaningful has begun. Old systems of exploitation and oppression are passing away and new systems of justice and equality are being born. In a real sense ours is a great time in which to be alive. Therefore I am not yet discouraged about the future. Granted that the easygoing optimism of yesterday is impossible. Granted that we face a world crisis which often leaves us standing amid the surging murmur of life's restless sea. But every crisis has both its dangers and its opportunities. Each can spell either salvation or doom. In a dark, confused world the spirit of God may yet reign supreme.

This article is a restatement of chapter 6 of Martin Luther King, Jr., *Stride Toward Freedom: The Montgomery Story* (New York: Harper & Row, 1958) which appeared in *Christian Century* 77 (April 13, 1960): 439–41.

8

The Rising Tide of Racial Consciousness

(1960)

A growing sense of racial pride and consciousness always accompanied African Americans who achieved success despite the discriminatory impediments white racists constructed. But many young black people were not prepared to be satisfied with the success of a few individuals. In fact this only raised questions about why so few could satisfy the standards of the system. Some racists began to restate an old question, "Are African people inherently inferior to white people?" Those in the older generation, such as Dr. Kenneth B. Clark, the famous black child psychologist, embraced the liberal belief that there can be no social equality where there is no equal opportunity to succeed. In fact this philosophy became the philosophical linchpin that bound those who successfully sued the Topeka, Kansas, Board of Education in 1954 for the Reverend Oliver Brown's right to send his black daughter to a predominantly white neighborhood school.

Mrs. Ella Baker, a staff member of the Southern Christian Leadership Conference, encouraged the young people who organized the Student Nonviolent Coordinating Committee (SNCC) to think for themselves. SNCC came into being between April 15 and 17, 1960, on the campus of Shaw University in Raleigh, North Carolina. This integrated, though predominantly black, group of young people believed that they had to take their future into their own hands. Meanwhile, black nationalism was also reasserting itself on college campuses and in the streets of many urban centers, especially northern ones. In fact on July 31, Elijah Muhammad, the head of the Nation of Islam, called upon African Americans to form a state of their own.

As this Black Consciousness Movement began to accelerate, Dr. King, who greatly sympathized with it, reminded middle-class African Americans that they need not be afraid of proclaiming racial pride. He also reminded advocates of Black Consciousness that racial pride should not encourage racial isolationism. As he later said in his "I Have A Dream" speech, all Americans hold the "promissory note" signed by the founders of the United States. He pleaded with all Americans not to allow the nation to default on its indebtedness to the spirit and principle of democracy. He saw the quest for racial justice to be an integral part of that larger struggle. The following address is an abridged version of a speech that Dr. King delivered before the Golden Anniversary Conference of the National Urban League.

WHAT ARE THE FACTORS THAT HAVE LED TO THIS NEW SENSE OF DIGNITY and self-respect on the part of the Negro? First, we must mention the population shift from rural to urban life. For many years the vast majority of Negroes were isolated on the rural plantation. They had very little contact with the world outside their geographical boundaries. But gradually circumstances made it possible and necessary for them to migrate to new and larger centers—the spread of the automobile, the Great Depression, and the social upheavals of the two world wars. These new contacts led to a broadened outlook. These new levels of communication brought new and different attitudes.

A second factor that has caused the Negroes' new self-consciousness has been rapid educational advance. Over the years there has been a steady decline of crippling illiteracy. At emancipation only five percent of the Negroes were literate; today more than ninety-five percent are literate. Constant streams of Negro students are finishing colleges and universities every year. More than sixteen hundred Negroes have received the highest academic degree bestowed by an

American university. These educational advances have naturally broadened his thinking. They have given the Negro not only a larger view of the world, but also a larger view of himself.

A third factor that produced the new sense of pride in the Negro was the gradual improvement of his economic status. While the Negro is still the victim of tragic economic exploitation, significant strides have been made. The annual collective income of the Negro is now approximately eighteen billion dollars, which is more than the national income of Canada and all of the exports of the United States. This augmented purchasing power has been reflected in more adequate housing, improved medical care, and greater educational opportunities. As these changes have taken place they have driven the Negro to change his image of himself.

A fourth factor that brought about the new sense of pride in the Negro was the Supreme Court's decision outlawing segregation in the public schools. For all men of good will May 17, 1954, came as a joyous daybreak to end the long night of enforced segregation. In simple, eloquent, and unequivocal language the court affirmed that "separate but equal" facilities are inherently unequal and that to segregate a child on the basis of his race is to deny that child equal protection of the law. This decision brought hope to millions of disinherited Negroes who had formerly dared only to dream of freedom. Like an exit sign that suddenly appeared to one who had walked through a long and desolate corridor, this decision came as a way out of the darkness of segregation. It served to transform the fatigue of despair into the buoyancy of hope. It further enhanced the Negro's sense of dignity.

A fifth factor that has accounted for the new sense of dignity on the part of the Negro has been the awareness that his struggle for freedom is a part of a worldwide struggle. He has watched developments in Asia and Africa with rapt attention. On these vast prodigious continents dwell two-thirds of the world's people. For years they were exploited economically, dominated politically, segregated and humiliated by foreign powers. Thirty years ago there were only three independent countries in the whole of Africa—Liberia, Ethiopia and South Africa. By 1962, there may be as many as thirty independent nations in Africa. These rapid changes have naturally influenced the thinking of the American Negro. He knows that his struggle for human dignity is not an isolated event. It is a drama being played on the stage of the world with spectators and supporters from every continent.

DETERMINATION AND RESISTANCE

This growing self-respect has inspired the Negro with a new determination to struggle and sacrifice until first-class citizenship becomes a reality. This is at bottom the meaning of what is happening in the South today. Whether it is manifested in nine brave children of Little Rock walking through jeering and hostile mobs, or fifty thousand people of Montgomery, Alabama, substituting tired feet for tired souls and walking the streets of that city for 381 days, or thousands of courageous students electrifying the nation by quietly and nonviolently sitting at lunch counters that have been closed to them because of the color of their skin, the motivation is always the same—the Negro would rather suffer in dignity than accept segregation in humiliation.

This new determination on the part of the Negro has not been welcomed by some segments of the nation's population. In some instances it has collided with tenacious and determined resistance. This resistance has risen at times to ominous proportions. A few states have reacted in open defiance. The legislative halls of the South ring loud with such words as "interposition" and "nullification." Many public officials are going to the absurd and fanatical extreme of closing the schools rather than to comply with the law of the land. This resistance to the Negroes' aspirations expresses itself in the resurgence of the Ku Klux Klan and the birth of White Citizens Councils.

The resistance to the Negroes' aspirations expresses itself not only in obvious methods of defiance, but in the subtle and skillful method of truth distortion. In an attempt to influence the minds of northern and southern liberals, the segregationists will cleverly disseminate half-truths. Instead of arguing for the validity of segregation and racial inferiority on the basis of the Bible, they set their arguments on cultural and sociological grounds. The Negro is not ready for integration, they say; because of academic and cultural lags on the part of the Negro, the integration of schools will pull the white race down. They are never honest enough to admit that the academic and cultural lags in the Negro community are themselves the result of segregation and discrimination. The best way to solve any problem is to remove the cause. It is both rationally unsound and sociologically untenable to use the tragic effects of segregation as an argument for its continuation.

The great challenge facing the nation today is to solve this pressing problem and bring into full realization the ideals and dreams of our democracy. How we deal with this crucial situation will determine our political health as a nation and our prestige as a leader of the free world. The price that America must pay for the continued oppression of the Negro is the price of its own destruction. The hour is late; the clock of destiny is ticking out. We must act now! It is a trite yet urgently true observation that if America is to remain a first-class nation, it cannot have second-class citizens.

Our primary reason for bringing an end to racial discrimination in America must not be the Communist challenge. Nor must it be merely to appeal to Asian and African peoples. The primary reason for our uprooting racial discrimination from our society is that it is morally wrong. It is a cancerous disease that prevents us from realizing the sublime principles of our Judeo-Christian tradition. Racial discrimination substitutes an "I-it" relationship for the "I-thou" relationship. It relegates persons to the status of things. Whenever racial discrimination exists it is a tragic expression of man's spiritual degeneracy and moral bankruptcy. Therefore, it must be removed not merely because it is diplomatically expedient, but because it is morally compelling.

A NATIONAL PROBLEM

The racial issue that we confront in America is not a sectional but a national problem. Injustice anywhere is a threat to justice everywhere. Therefore, no American can afford to be apathetic about the problem of racial justice. It is a problem that meets every man at his front door.

There is need for strong and aggressive leadership from the federal government. There is a pressing need for a liberalism in the North that is truly liberal, that firmly believes in integration in its own community as well as in the deep South. There is need for the type of liberal who not only rises up with righteous indignation when a Negro is lynched in Mississippi, but will be equally incensed when a Negro is denied the right to live in his neighborhood, or join his professional association, or secure a top position in his business. This is no day to pay mere lip service to integration; we must pay life service to it.

There are several other agencies and groups that have significant roles to play in this all-important period of our nation's history; the problem of racial injustice is so weighty in detail and broad in extent that it requires the concerted efforts of numerous individuals and institutions to bring about a solution.

THE PRIMARY RESPONSIBILITY

In the final analysis if first-class citizenship is to become a reality for the Negro he must assume the primary responsibility for making it so. The Negro must not be victimized with the delusion of thinking that others should be more concerned than himself about his citizenship rights.

In this period of social change the Negro must work on two fronts. On the one hand we must continue to break down the barrier of segregation. We must resist all forms of racial injustice. This resistance must always be on the highest level of dignity and discipline. It must never degenerate to the crippling level of violence. There is another way—a way as old as the insights of Jesus of Nazareth and as modern as the methods of Mahatma Gandhi. It is a way not for the weak and cowardly but for the strong and courageous. It has been variously called passive resistance, nonviolent resistance, or simply Christian love. It is my great hope that, as the Negro plunges deeper into the quest for freedom, he will plunge deeper into the philosophy of nonviolence. As a race we must work passionately and unrelentingly for first-class citizenship, but we must never use second-class methods to gain it. Our aim must be not to defeat or humiliate the white man, but to win his friendship and understanding. We must never become bitter nor should we succumb to the temptation of using violence in the struggle, for if this happens, unborn generations will be the recipients of a long and desolate night of bitterness and our chief legacy to the future will be an endless reign of meaningless chaos.

I feel that this way of nonviolence is vital because it is the only way to reestablish the broken community. It is the method which seeks to implement the just law by appealing to the conscience of the great decent majority who through blindness, fear, pride, or irrationality have allowed their consciences to sleep.

The nonviolent resisters can summarize their message in the following simple terms: we will take direct action against injustice without waiting for other agencies to act. We will not obey unjust laws or submit to unjust practices. We will do this peacefully, openly, and cheerfully because our aim is to persuade. We adopt the means of nonviolence because our end is a community at peace with itself. We will try to persuade with our words, but, if our words fail, we will try to persuade with our acts. We will always be willing to talk and seek fair compromise, but we are ready to suffer when necessary and even risk our lives to become witnesses to the truth as we see it.

I realize that this approach will mean suffering and sacrifice. It may mean going to jail. If such is the case the resister must be willing to fill the jail houses of the South. It may even mean physical death. But if physical death is the price that a man must pay to free his children and his white brethren from a permanent death of the spirit, then nothing could be more redemptive. This is the type of soul force that I am convinced will triumph over the physical force of the oppressor.

This approach to the problem of oppression is not without successful precedent. We have the magnificent example of Gandhi who challenged the might of the British Empire and won independence for his people by using only the weapons of truth, noninjury, courage, and soul force. Today we have the example of thousands of Negro students in the South who have courageously challenged the principalities of segregation. These young students have taken the deep groans and the passionate yearnings of the Negro people and filtered them in their own souls and fashioned them in a creative protest which is an epic known all over our nation. For the last few months they have moved in a uniquely meaningful orbit imparting light and heat to distant satellites. Through their nonviolent direct action they have been able to open hundreds of formerly segregated lunch counters in almost eighty cities. It is no overstatement to characterize these events as historic. Never before in the United States has so large a body of students spread a struggle over so great an area in pursuit of a goal of human dignity and freedom. I am convinced that future historians will have to record this student movement as one of the greatest epics of our heritage.

Let me mention another front on which we must work that is equally significant. The Negro must make a vigorous effort to improve his personal standards. The only answer that we can give to those who through blindness and fear would question our readiness and capability is that our lagging standards exist because of the legacy of slavery and segregation, inferior schools, slums, and second-class citizenship, and not because of an inherent inferiority. The fact that so many Negroes have made lasting and significant contributions to the cultural life of America in spite of these crippling restrictions is sufficient to refute all of the myths and half-truths disseminated by the segregationist.

Yet we cannot ignore the fact that our standards do often fall short. One of the sure signs of maturity is the ability to rise to the point of self-criticism. We have been affected by our years of economic deprivation and social isolation. Some Negroes have become cynical and disillusioned. Some have so conditioned themselves to the system of segregation that they have lost that creative something called *initiative*. So many have used their oppression as an excuse for mediocrity. Many of us live above our means, spend money on nonessentials and frivolities, and fail to give to serious causes, organizations, and educational institutions that so desperately need funds. Our crime rate is far too high.

CONSTRUCTIVE ACTION

Therefore there is a pressing need for the Negro to develop a positive program through which these standards can be improved. After we have analyzed the sociological and psychological causes of these problems, we must seek to develop a constructive program to solve them. We must constantly stimulate our youth to rise above the stagnant level of mediocrity and seek to achieve excellence in their various fields of endeavor. Doors are opening now that were not open in the past, and the great challenge facing minority groups is to be ready to enter these doors as they open. No greater tragedy could befall us at this hour but that of allowing new opportunities to emerge without the concomitant preparedness to meet them.

We must make it clear to our young people that this is an age in which they will be forced to compete with people of all races and nationalities. We cannot aim merely to be good Negro teachers, good Negro doctors, or good Negro skilled laborers. We must set out to do a good job irrespective of race. We must seek to do our life's work so well that nobody could do it better. The Negro who seeks to be merely a good Negro, whatever he is, has already flunked his matriculation examination for entrance into the university of integration.

This then must be our present program: nonviolent resistance to all forms of racial injustice, even when this means going to jail; and bold, constructive action to end the demoralization caused by the legacy of slavery and segregation. The nonviolent struggle, if conducted with the dignity and courage already shown by the sit-in students of the South, will in itself help end the demoralization; but a new frontal assault on the poverty, disease, and ignorance of a people too long deprived of the God-given rights of life, liberty, and the pursuit of happiness will make the victory more certain.

We must work assiduously and with determined boldness to remove from the body politic this cancerous disease of discrimination which is preventing our democratic and Christian health from being realized. Then and only then will we be able to bring into full realization the dream of our American democracy—a dream yet unfulfilled. A dream of equality of opportunity, of privilege and property widely distributed; a dream of a land where men will not take necessities from the many to give luxuries to the few; a dream of a land where men do not argue that the color of a man's skin determines the content of his character; a dream of a place where all our gifts and resources are held not for ourselves alone but as instruments of service for the rest of humanity; the dream of a country where every man will respect the dignity and worth of all human personality, and men will dare to live together as brothers—that is the dream. Whenever it is fulfilled we will emerge from the bleak and desolate midnight of man's inhumanity to man into the bright and glowing daybreak of freedom and justice for all of God's children.

YWCA Magazine (December 1960): 4–6.

9

The Time for Freedom Has Come

(1961)

In an attempt to build on the success of Montgomery, many people, especially young people, embraced different strategies to confront segregation. The so-called "Freedom Ride" movement was the next major effort to undo segregationist laws that had gained the public spotlight after the Montgomery Bus Boycott (1955–1956) and the Sit-In movements (1960–1961). This time nonviolent resisters challenged laws that prohibited interracial seating on buses, trains, and airplanes crossing state borders. In the twentieth century, those who openly confronted laws and customs that prohibited blacks and whites from sitting next to each other on these public vehicles called themselves "Freedom Riders." But they had at least one predecessor in the nineteenth century.

In April 1885, T. McCants Stewart became the first black person to challenge segregation on interstate transportation vehicles by travelling through the South as if Jim Crow laws did not exist. He did not run into much difficulty since he was a lone traveler. But on April 9, 1947, the Congress on Racial Equality (CORE) sponsored the first integrated group of "Freedom Riders." This proved to be far more difficult than Stewart's effort because it was a direct challenge to Jim Crow. Bayard Rustin, who later served as one of Dr. King's chief strategists, was among this group. His experience on these rides, as well as his participation in CORE's successful attempt to dethrone Jim Crow in the downtown stores of St. Louis and Baltimore, was excellent preparation for the crucial advice he was able to offer Dr. King. Fourteen years later, on May 4, 1961, thirteen "Freedom Riders," including James Farmer, CORE's national director, boarded a bus in Washington, D.C., that was bound for the South. This widely publicized new phase of the Civil Rights Movement met violent opposition from white segregationists. Farmer's bus was bombed and burned just outside of Anniston, Alabama. The Freedom Riders themselves were attacked in Anniston, Birmingham, and Montgomery, Alabama, between May 14 and 20. President John F. Kennedy ordered his brother, Attorney General Robert F. Kennedy, to dispatch 400 federal marshals. A strenuous debate about who had final jurisdiction to maintain law and order ensued between the federal and state government. But rather than arrest the vigilantes, the local authorities arrested the Freedom Riders.

Meanwhile, the "pray-in" movement involved civil rights activists who sought to challenge religious segregation by forming small groups who would stage prayer vigils either inside a worship service or on the front steps of churches. A "wade-in" demonstration was even held at Rainbow Beach in Chicago on July 9 and July 16 when civil rights activists wore bathing suits on beaches that either legally or by custom excluded nonwhite people from using them. Local authorities would often arrest the demonstrators for disturbing the peace rather than detain their antagonists. A new generation of black youth became so committed to the cause for black freedom that they courageously went to jail to demonstrate their determination. Many abandoned their middle-class disdain for civil disobedience. They came to view arrest for the sake of liberation as a mark of honor. Dr. King here explains why that was the case.

ON A CHILL MORNING IN THE AUTUMN OF 1956, AN ELDERLY, TOILWORN Negro woman in Montgomery, Alabama, began her slow, painful four-mile walk to her job. It was the tenth month of the Montgomery bus boycott, which had begun with a life expectancy of one week. The old woman's difficult progress led a passerby to inquire sympathetically if her feet were tired. Her simple answer became the boycotter's watchword. "Yes, friend, my feet is real tired, but my soul is rested."

Five years passed and once more Montgomery arrested the world's attention. Now the symbolic segregationist is not a stubborn, rude bus driver. He emerges in 1961 as a hoodlum stomping the bleeding face of a freedom rider. But neither is the Negro today an elderly woman whose grammar is uncertain; rather, he is college-bred, Ivy League–clad, youthful, articulate and resolute. He has the imagination and drive of the young, tamed by discipline and commitment. The nation and the world have reacted with astonishment at these students cast from a new mold, unaware that a chain reaction was accumulating explosive force behind a strangely different facade.

Generating these changes is a phenomenon Victor Hugo described in these words: "There is no greater power on earth than an idea whose time has come." In the decade of the sixties the time for freedom for the Negro has come. This simple truth illuminates the motivations, the tactics and the objectives of the students' daring and imaginative movement.

The young Negro is not in revolt, as some have suggested, against a single pattern of timid, fumbling, conservative leadership. Nor is his conduct to be explained in terms of youth's excesses. He is carrying forward a revolutionary destiny of a whole people consciously and deliberately. Hence the extraordinary willingness to fill the jails as if they were honors classes and the boldness to absorb brutality, even to the point of death, and remain nonviolent. His inner strength derives from his goal of freedom and the leadership role he has grasped even at a time when some of his white counterparts still grope in philosophical confusion searching for a personal goal with human values, searching for security from economic instability, and seeking relief from the haunting fear of nuclear destruction.

The campuses of Negro colleges are infused with a dynamism of both action and philosophical discussion. The needs of a surging period of change have had an impact on all Negro groups, sweeping away conventional trivialities and escapism.

Even in the thirties, when the college campus was alive with social thought, only a minority was involved in action. Now, during the sit-in phase, when a few students were suspended or expelled, more than one college saw the total student body involved in a walk-out protest. This is a change in student activity of profound significance. Seldom, if ever, in American history has a student movement engulfed the whole student body of a college.

In another dimension, an equally striking change is altering the Negro campuses. Not long ago the Negro collegian imitated the white collegian. In attire, in athletics, in social life, imitation was the rule. For the future, he looked to a professional life cast in the image of the middle-class white professional. He imitated with such energy that Gunnar Myrdal described the ambitious Negro as "an exaggerated American."

Today the imitation has ceased. The Negro collegian now initiates. Groping for unique forms of protest, he created the sit-ins and freedom rides. Overnight his white fellow students began to imitate him. As the movement took hold, a revival of social awareness spread across campuses from Cambridge to California. It spilled over the boundaries of the single issue of desegregation and encompassed questions of peace, civil liberties, capital punishment and others. It penetrated the ivy-covered walls of the traditional institutions as well as the glass and stainless-steel structures of the newly established colleges.

A consciousness of leadership, a sense of destiny have given maturity and dedication to this generation of Negro students which have few precedents. As a minister, I am often given promises of dedication. Instinctively I examine the degree of sincerity. The striking quality in Negro students I have met is the intensity and depth of their commitment. I am no longer surprised to meet attractive, stylishly dressed young girls whose charm and personality would grace a junior prom and to hear them declare in unmistakably sincere terms, "Dr. King, I am ready to die if I must."

Many of the students, when pressed to express their inner feelings, identify themselves with students in Africa, Asia and South America. The liberation struggle in Africa has been the greatest single international influence on American Negro students. Frequently I hear them say that if their African brothers can break the bonds of colonialism, surely the American Negro can break Jim Crow.

African leaders such as President Kwame Nkrumah of Ghana, Governor General Nnamdi Azikiwe of Nigeria, Dr. Tom Mboya of Kenya and Dr. Hastings Banda of Nyasaland are popular heroes on most Negro college campuses. Many groups demonstrated or otherwise protested when the Congo leader, Patrice Lumumba, was assassinated. The newspapers were mistaken when they interpreted these outbursts of indignation as "Communist-inspired."

Part of the impatience of Negro youth stems from their observation that change is taking place rapidly in Africa and other parts of the world, but comparatively slowly in the South. When the United States Supreme Court handed down its historic desegregation decision in 1954, many of us, perhaps naively, thought that great and sweeping school integration would ensue. Yet, today, seven years later, only seven percent of the Negro children of the South have been placed in desegregated schools. At the current rate it will take ninety-three more years to desegregate the public schools of the South. The collegians say, "We can't wait that long" or simply, "We won't wait!"

Negro students are coming to understand that education and learning have become tools for shaping the future and not devices of privilege for an exclusive few. Behind this spiritual explosion is the shattering of a material atom.

The future of the Negro college student has long been locked within the narrow walls of limited opportunity. Only a few professions could be practiced by Negroes and, but for a few exceptions, behind barriers of segregation in the North as well as the South. Few frustrations can compare with the experience of struggling with complex academic subjects, straining to absorb concepts which may never be used, or only half-utilized under conditions insulting to the trained mind. A Negro intern blurted out to me shortly after his patient died, "I wish I were not so well trained because then I would never know how many of these people need not die for lack of proper equipment, adequate post-operative care and timely admission. I'm not practicing good medicine. I'm presiding over tragedies which the absence of good medicine creates."

The Negro lawyer knows his practice will bulk large with criminal cases. The law of wills, of corporations, of taxation will only infrequently reach his office because the clientele he serves has had little opportunity to accumulate property. In his courtroom experience in the South, his clients and witnesses will probably be segregated and he, as well as they, will seldom be referred to as "Mister." Even worse, all too often he knows that the verdict was sealed the moment the arrest was made.

These are but a few examples of the real experience of the Negro professional, seen clearly by the student who has been asked to study with serious purpose. Obviously his incentive has been smothered and weakened. But today, more than ever, the Negro realizes that, while studying, he can also act to change the conditions which cripple his future. In the struggle to desegregate society he is altering it directly for himself as well as for future generations.

There is another respect in which the Negro student is benefiting, and simultaneously contributing to, society as a whole. He is learning social responsibility; he is learning to earn, through his own direct sacrifice, the result he seeks. There are those who would make him soft, pliable and conformist—a mechanical organization man or an uncreative status seeker. But the experience of Negro youth is as harsh and demanding as that of the pioneer on the untamed frontier. Because his struggle is complex, there is no place in it for the frivolous or rowdy. Knowledge and discipline are as indispensable as courage and self-sacrifice. Hence the forging of priceless qualities of character is taking place daily as a high moral goal is pursued.

Inevitably there will emerge from this caldron a mature man, experienced in life's lessons, socially aware, unafraid of experimentation and, most of all, imbued with the spirit of service and dedication to a great ideal. The movement therefore gives to its participants a double education—academic learning from books and classes, and life's lessons from responsible participation in social action. Indeed, the answer to the quest for a more mature, educated American, to compete successfully with the young people of other lands, may be present in this new movement.

Of course, not every student in our struggle has gained from it. This would be more than any humanly designed plan could realize. For some, the opportunity for personal advantage presented itself and their character was not equal to the challenge. A small percentage of students have found it convenient to escape from their own inadequacies by identifying with the sit-ins and other activities. They are, however, relatively few because this is a form of escape in which the flight from responsibility imposes even greater responsibilities and risks.

It is not a solemn life, for all of its seriousness. During a vigorous debate among a group of students discussing the moral and practical soundness of nonviolence, a majority rejected the employment of force. As the minority dwindled to a single student, he finally declared, "All I know is that, if rabbits could throw rocks, there would be fewer hunters in the forest."

This is more than a witty remark to relieve the tensions of serious and even grim discussion. It expresses some of the pent-up impatience, some of the discontent and some of the despair produced by minute corrections in the face of enormous evil. Students necessarily have conflicting reactions. It is understandable that violence presents itself as a quick, effective answer for a few.

For the large majority, however, nonviolent, direct action has emerged as the better and more successful way out. It does not require that they abandon their discontent. This discontent is a sound, healthy social response to the injustice and brutality they see around them. Nonviolence offers a method by which they can fight the evil with which they cannot live. It offers a unique weapon which, without firing a single bullet, disarms the adversary. It exposes his moral defenses, weakens his morale, and at the same time works on his conscience.

Another weapon which Negro students have employed creatively in their nonviolent struggle is satire. It has enabled them to avoid corrosive anger while pressing the cutting edge of ridicule against the opponent. When they have been admonished to "go slow," patiently to wait for gradual change, with a straight face they will assure you that they are diligently searching for the happy medium between the two extremes of moderation and gradualism.

It is perhaps the special quality of nonviolent direct action, which sublimates anger, that explains why so few students are attracted to extreme nationalist sects advocating black supremacy. The students have anger under controlling bonds of discipline. Hence they can answer appeals for cooling-off periods by advocating cooling-off for those who are hot with anger and violence.

Much has been made of the willingness of these devotees of nonviolent social action to break the law. Paradoxically, although they have embraced Thoreau's and Gandhi's civil disobedience on a scale dwarfing any past experience in American history, they do respect law. They feel a moral responsibility to obey just laws. But they recognize that there are also unjust laws.

From a purely moral point of view, an unjust law is one that is out of harmony with the moral law of the universe. More concretely, an unjust law is one in which the minority is compelled to observe a code that is not binding on the majority. An unjust law is one in which people are required to obey a code that they had no part in making because they were denied the right to vote.

In disobeying such unjust laws, the students do so peacefully, openly and nonviolently. Most important, they willingly accept the penalty, whatever it is, for in this way the public comes to reexamine the law in question and will thus decide whether it uplifts or degrades man.

This distinguishes their position on civil disobedience from the "uncivil disobedience" of the segregationist. In the face of laws they consider unjust, the racists seek to defy, evade and circumvent the law, and they are unwilling to accept the penalty. The end result of their defiance is anarchy and disrespect for the law. The students, on the other hand, believe that he who openly disobeys a law, a law conscience tells him is unjust, and then willingly accepts the penalty, gives evidence thereby that he so respects that law that he belongs in jail until it is changed. Their appeal is to the conscience.

Beyond this, the students appear to have perceived what an older generation overlooked in the role of law. The law tends to declare rights—it does not deliver them. A catalyst is needed to breathe life experience into a judicial decision by the persistent exercise of the rights until they become usual and ordinary in human conduct. They have offered their energies, their bodies to effect this result. They see themselves the obstetricians at the birth of a new order. It is in this manner that the students have related themselves to and materialized "the idea whose time has come."

In a sense, the victories of the past two years have been spectacular and considerable. Because of the student sitters, more than 150 cities in the South have integrated their lunch counters. Actually, the current breakthroughs have come about partly as a result of the patient legal, civil and social ground clearing of the previous decades. Then, too, but slowly, the national government is realizing that our so-called domestic race relations are a major force in our foreign relations. Our image abroad reflects our behavior at home.

Many liberals, of the North as well as the South, when they list the unprecedented programs of the past few years, yearn for a "cooling-off" period; not too fast, they say, we may lose all that we have gained if we push faster than the violent ones can be persuaded to yield.

This view, though understandable, is a misreading of the goals of the young Negroes. They are not after "mere tokens" of integration ("tokenism," they call it); rather theirs is a revolt against the whole system of Jim Crow and they are prepared to sit-in, kneel-in, wade-in and stand-in until every waiting room, rest room, theatre and other facility throughout the nation that is supposedly open to the public is in fact open to Negroes, Mexicans, Indians, Jews or what have you. Theirs is total commitment to this goal of equality and dignity. And for this achievement they are prepared to pay the costs—whatever they are—in suffering and hardship as long as may be necessary.

Indeed, these students are not struggling for themselves alone. They are seeking to save the soul of America. They are taking our whole nation back to those great wells of democracy which were dug deep by the Founding Fathers in the formulation of the Constitution and the Declaration of Independence. In sitting down at the lunch counters, they are in reality standing up for the best in the American dream. They courageously go to the jails of the South in order to get America out of the dilemma in which she finds herself as a result of the continued existence of segregation. One day historians will record this student movement as one of the most significant epics of our heritage.

But should we, as a nation, sit by as spectators when the social unrest seethes? Most of us recognize that the Jim Crow system is doomed. If so, would it not be the wise and human thing to abolish the system surely and swiftly? This would not be difficult, if our national government would exercise its full powers to enforce federal laws and court decisions and do so on a scale commensurate with the problems and with an unmistakable decisiveness. Moreover, we would need our religious, civic and economic leaders to mobilize their forces behind a real, honest-to-goodness "End Jim Crow Now" campaign.

This is the challenge of these young people to us and our ideals. It is also an expression of their new-found faith in themselves as well as in their fellow man.

In an effort to understand the students and to help them understand themselves, I asked one student I know to find a quotation expressing his feeling of our struggle. He was an inarticulate young man, athletically expert and far more poetic with a basketball than with words, but few would have found the quotation he typed on a card and left on my desk early one morning:

> I sought my soul, but my soul I could not see,
> I sought my God, but he eluded me,
> I sought my brother, and I found all three.

New York Times Magazine (September 10, 1961): 25ff.

10

Letter from a Birmingham Jail

(1963)

Friendly detractors had accused Martin Luther King, Jr., of urging others to do things that he did not do. Some self-righteously proclaimed that Dr. King did not spend enough time in jail. It hurt him greatly to receive such criticisms from some of his Student Nonviolent Coordinating Committee colleagues after their 1962 voter registration campaign in Albany, Georgia. Such petty criticisms resulted more from envy and forgetfulness than from actual facts. And, they seldom took his leadership responsibilities into account.

His critics did not know that the Southern Christian Leadership Conference's cash flow problems often placed it on the edge of financial ruin. Monies from Dr. King's speaking engagements frequently saved SCLC from this kind of embarrassment. Furthermore, a simple catalogue of the many times he was arrested would cast great doubts on the accuracy of such criticisms. The many hours he spent celebrating "the sacrament of imprisonment" became spiritual preparation for the "Letter from a Birmingham Jail." He wrote this essay in the form of an open letter on April 16, 1963, while he was serving a sentence for participating in civil rights demonstrations in Birmingham, Alabama.

Dr. King rarely took time to defend himself against his opponents. But eight prominent "liberal" Alabama clergymen published an open letter earlier in January that called on King to allow the battle for integration to continue in the local and federal courts. They warned that King's nonviolent resistance would have the effect of inciting civil disturbances. Dr. King wanted Christian ministers to see that the meaning of Christian discipleship was at the heart of the African American struggle for freedom, justice and equality.

MY DEAR FELLOW CLERGYMEN,

While confined here in the Birmingham city jail, I came across your recent statement calling our present activities "unwise and untimely." Seldom, if ever, do I pause to answer criticism of my work and ideas. If I sought to answer all of the criticisms that cross my desk, my secretaries would be engaged in little else in the course of the day, and I would have no time for constructive work. But since I feel that you are men of genuine good will and your criticisms are sincerely set forth, I would like to answer your statement in what I hope will be patient and reasonable terms.

I think I should give the reason for my being in Birmingham, since you have been influenced by the argument of "outsiders coming in." I have the honor of serving as president of the Southern Christian Leadership Conference, an organization operating in every southern state, with headquarters in Atlanta, Georgia. We have some eighty-five affiliate organizations all across the South—one being the Alabama Christian Movement for Human Rights. Whenever necessary and possible we share staff, educational and financial resources with our affiliates. Several months ago our local affiliate here in Birmingham invited us to be on call to engage in a nonviolent direct-action program if such were deemed necessary. We readily consented and when the hour came we lived up to our promises. So I am here, along with several members of my staff, because we were invited here. I am here because I have basic organizational ties here.

Beyond this, I am in Birmingham because injustice is here. Just as the eighth-century prophets left their little villages and carried their "thus saith the Lord" far beyond the boundaries of their hometowns; and just as the Apostle Paul left his little village of Tarsus and carried the gospel of Jesus Christ to practically every hamlet and city of the Graeco-Roman world, I too am compelled to carry the gospel of freedom beyond my particular hometown. Like Paul, I must constantly respond to the Macedonian call for aid.

Moreover, I am cognizant of the interrelatedness of all communities and states. I cannot sit idly by in Atlanta and not be concerned about what happens in Birmingham. Injustice anywhere is a threat to justice everywhere. We are caught in an inescapable network of mutuality, tied in a single garment of destiny. Whatever affects one directly affects all indirectly. Never again can we afford to live with the narrow, provincial "outside agitator" idea. Anyone who lives in the United States can never be considered an outsider anywhere in this country.

You deplore the demonstrations that are presently taking place in Birmingham. But I am sorry that your statement did not express a similar concern for the conditions that brought the demonstrations into being. I am sure that each of you would want to go beyond the superficial social analyst who looks merely at effects, and does not grapple with underlying causes. I would not hesitate to say that it is unfortunate that so-called demonstrations are taking place in Birmingham at this time, but I would say in more emphatic terms that it is even more unfortunate that the white power structure of this city left the Negro community with no other alternative.

In any nonviolent campaign there are four basic steps: (1) collection of the facts to determine whether injustices are alive, (2) negotiation, (3) self-purification, and (4) direct action. We have gone through all of these steps in Birmingham. There can be no gainsaying of the fact that racial injustice engulfs this community.

Birmingham is probably the most thoroughly segregated city in the United States. Its ugly record of police brutality is known in every section of this country. Its unjust treatment of Negroes in the courts is a notorious reality. There have been more unsolved bombings of Negro homes and churches in Birmingham than any city in this nation. These are the hard, brutal and unbelievable facts. On the basis of these conditions Negro leaders sought to negotiate with the city fathers. But the political leaders consistently refused to engage in good faith negotiation.

Then came the opportunity last September to talk with some of the leaders of the economic community. In these negotiating sessions certain promises were made by the merchants—such as the promise to remove the humiliating racial signs from the stores. On the basis of these promises Rev. Shuttlesworth and the leaders of the Alabama Christian Movement for Human Rights agreed to call a moratorium on any type of demonstrations. As the weeks and months unfolded we realized that we were the victims of a broken promise. The signs remained. Like so many experiences of the past we were confronted with blasted hopes, and the dark shadow of a deep disappointment settled upon us. So we had no alternative except that of preparing for direct action, whereby we would present our very bodies as a means of laying our case before the conscience of the local and national community. We were not unmindful of the difficulties involved. So we decided to go through a process of self-purification. We started having workshops on nonviolence and repeatedly asked ourselves the questions, "Are you able to accept blows without retaliating?" "Are you able to endure the ordeals of jail?" We decided to set our direct-action program around the Easter season, realizing that with the exception of Christmas, this was the largest shopping period of the year. Knowing that a strong economic withdrawal program would be the by-product of direct action, we felt that this was the best time to bring pressure on the merchants for the needed changes. Then it occurred to us that the March election was ahead and so we speedily decided to postpone action until after election day. When we discovered that Mr. Connor was in the run-off, we decided again to postpone action so that the demonstrations could not be used to cloud the issues. At this time we agreed to begin our nonviolent witness the day after the run-off.

This reveals that we did not move irresponsibly into direct action. We too wanted to see Mr. Connor defeated; so we went through postponement after postponement to aid in this community need. After this we felt that direct action could be delayed no longer.

You may well ask, "Why direct action? Why sit-ins, marches, etc.? Isn't negotiation a better path?" You are exactly right in your call for negotiation. Indeed, this is the purpose of direct action. Nonviolent direct action seeks to create such a crisis and establish such creative tension that a community that has constantly refused to negotiate is forced to confront the issue. It seeks so to dramatize the issue that it can no longer be ignored. I just referred to the creation of tension as a part of the work of the nonviolent resister. This may sound rather shocking. But I must confess that I am not afraid of the word *tension*. I have earnestly worked and preached against violent tension, but there is a type of constructive nonviolent tension that is necessary for growth. Just as Socrates felt that it was necessary to create a tension in the mind so that individuals could rise from the bondage of myths and half-truths to the unfettered realm of creative analysis and objective appraisal, we must see the need of having nonviolent gadflies to create the kind of tension in society that will help men to rise from the dark depths of prejudice and racism to the majestic heights of understanding and brotherhood. So the purpose of the direct action is to create a situation so crisis-packed that it will inevitably open the door to negotiation. We, therefore, concur with you in your call for negotiation. Too long has our beloved Southland been bogged down in the tragic attempt to live in monologue rather than dialogue.

One of the basic points in your statement is that our acts are untimely. Some have asked, "Why didn't you give the new administration time to act?" The only answer that I can give to this inquiry is that the new administration must be prodded about as much as the outgoing one before it acts. We will be sadly mistaken if we feel that the election of Mr. Boutwell will bring the millennium to Birmingham. While Mr. Boutwell is much more articulate and gentle than Mr. Connor, they are both segregationists, dedicated to the task of maintaining the status quo. The hope I see in Mr. Boutwell is that he will be reasonable enough to see the futility of massive resistance to desegregation. But he will not see this without pressure from the devotees of civil rights. My friends, I must say to you that we have not made a single gain in civil rights without determined legal and nonviolent pressure. History is the long and tragic story of the fact that privileged groups seldom give up their privileges voluntarily. Individuals may see the moral light and voluntarily give up their unjust posture; but as Reinhold Niebuhr has reminded us, groups are more immoral than individuals.

We know through painful experience that freedom is never voluntarily given by the oppressor; it must be demanded by the oppressed. Frankly, I have never yet engaged in a direct action movement that was "well-timed," according to the timetable of those who have not suffered unduly from the disease of segregation. For years now I have heard the word "Wait!" It rings in the ear of every

Negro with a piercing familiarity. This "Wait" has almost always meant "Never." It has been a tranquilizing thalidomide, relieving the emotional stress for a moment, only to give birth to an ill-formed infant of frustration. We must come to see with the distinguished jurist of yesterday that "justice too long delayed is justice denied." We have waited for more than 340 years for our constitutional and God-given rights. The nations of Asia and Africa are moving with jet-like speed toward the goal of political independence, and we still creep at horse and buggy pace toward the gaining of a cup of coffee at a lunch counter. I guess it is easy for those who have never felt the stinging darts of segregation to say, "Wait." But when you have seen vicious mobs lynch your mothers and fathers at will and drown your sisters and brothers at whim; when you have seen hate-filled policemen curse, kick, brutalize and even kill your black brothers and sisters with impunity; when you see the vast majority of your twenty million Negro brothers smothering in an airtight cage of poverty in the midst of an affluent society; when you suddenly find your tongue twisted and your speech stammering as you seek to explain to your six-year-old daughter why she can't go to the public amusement park that has just been advertised on television, and see tears welling up in her little eyes when she is told that Funtown is closed to colored children, and see the depressing clouds of inferiority begin to form in her little mental sky, and see her begin to distort her little personality by unconsciously developing a bitterness toward white people; when you have to concoct an answer for a five-year-old son asking in agonizing pathos: "Daddy, why do white people treat colored people so mean?"; when you take a cross-country drive and find it necessary to sleep night after night in the uncomfortable corners of your automobile because no motel will accept you; when you are humiliated day in and day out by nagging signs reading "white" and "colored"; when your first name becomes "nigger" and your middle name becomes "boy" (however old you are) and your last name becomes "John," and when your wife and mother are never given the respected title "Mrs."; when you are harried by day and haunted by night by the fact that you are a Negro, living constantly at tiptoe stance never quite knowing what to expect next, and plagued with inner fears and outer resentments; when you are forever fighting a degenerating sense of "nobodiness"; then you will understand why we find it difficult to wait. There comes a time when the cup of endurance runs over, and men are no longer willing to be plunged into an abyss of injustice where they experience the blackness of corroding despair. I hope, sirs, you can understand our legitimate and unavoidable impatience.

You express a great deal of anxiety over our willingness to break laws. This is certainly a legitimate concern. Since we so diligently urge people to obey the Supreme Court's decision of 1954 outlawing segregation in the public schools, it is rather strange and paradoxical to find us consciously breaking laws. One may well ask, "How can you advocate breaking some laws and obeying others?" The answer is found in the fact that there are two types of laws: there are *just* and there are *unjust* laws. I would agree with Saint Augustine that "An unjust law is no law at all."

Now what is the difference between the two? How does one determine when a law is just or unjust? A just law is a man-made code that squares with the moral law or the law of God. An unjust law is a code that is out of harmony with the moral law. To put it in the terms of Saint Thomas Aquinas, an unjust law is a human law that is not rooted in eternal and natural law. Any law that uplifts human personality is just. Any law that degrades human personality is unjust. All segregation statutes are unjust because segregation distorts the soul and damages the personality. It gives the segregator a false sense of superiority, and the segregated a false sense of inferiority. To use the words of Martin Buber, the great Jewish philosopher, segregation substitutes an "I-it" relationship for the "I-thou" relationship, and ends up relegating persons to the status of things. So segregation is not only politically, economically and sociologically unsound, but it is morally wrong and sinful. Paul Tillich has said that sin is separation. Isn't segregation an existential expression of man's tragic separation, an expression of his awful estrangement, his terrible sinfulness? So I can urge men to disobey segregation ordinances because they are morally wrong.

Let us turn to a more concrete example of just and unjust laws. An unjust law is a code that a majority inflicts on a minority that is not binding on itself. This is difference made legal. On the other hand a just law is a code that a majority compels a minority to follow that it is willing to follow itself. This is sameness made legal.

Let me give another explanation. An unjust law is a code inflicted upon a minority which that minority had no part in enacting or creating because they did not have the unhampered right to vote. Who can say that the legislature of Alabama which set up the segregation laws was democratically elected? Throughout the state of Alabama all types of conniving methods are used to prevent Negroes from becoming registered voters and there are some counties without a single Negro registered to vote despite the fact that the Negro constitutes a majority of the population. Can any law set up in such a state be considered democratically structured?

These are just a few examples of unjust and just laws. There are some instances when a law is just on its face and unjust in its application. For instance, I was arrested Friday on a change of parading without a permit. Now there is nothing wrong with an ordinance which requires a permit for a parade, but when the ordinance is used to preserve segregation and to deny citizens the First Amendment privilege of peaceful assembly and peaceful protest, then it becomes unjust.

I hope you can see the distinction I am trying to point out. In no sense do I advocate evading or defying the law as the rabid segregationist would do. This would lead to anarchy. One who breaks an unjust law must do it *openly, lovingly* (not hatefully as the white mothers did in New Orleans when they were seen on television screaming, "nigger, nigger, nigger"), and with a willingness to accept the penalty. I submit that an individual who breaks a law that conscience tells him is unjust, and willingly accepts the penalty by staying in jail to arouse the conscience of the community over its injustice, is in reality expressing the very highest respect for law.

Of course, there is nothing new about this kind of civil disobedience. It was seen sublimely in the refusal of Shadrach, Meshach and Abednego to obey the laws of Nebuchadnezzar because a higher moral law was involved. It was practiced superbly by the early Christians who were willing to face hungry lions and the excruciating pain of chopping blocks, before submitting to certain unjust laws of the Roman Empire. To a degree academic freedom is a reality today because Socrates practiced civil disobedience.

We can never forget that everything Hitler did in Germany was "legal" and everything the Hungarian freedom fighters did in Hungary was "illegal." It was "illegal" to aid and comfort a Jew in Hitler's

Germany. But I am sure that if I had lived in Germany during that time I would have aided and comforted my Jewish brothers even though it was illegal. If I lived in a Communist country today where certain principles dear to the Christian faith are suppressed, I believe I would openly advocate disobeying these anti-religious laws. I must make two honest confessions to you, my Christian and Jewish brothers. First, I must confess that over the last few years I have been gravely disappointed with the white moderate. I have almost reached the regrettable conclusion that the Negro's great stumbling block in the stride toward freedom is not the White Citizens Counciler or the Ku Klux Klanner, but the white moderate who is more devoted to "order" than to justice; who prefers a negative peace which is the absence of tension to a positive peace which is the presence of justice; who constantly says, "I agree with you in the goal you seek, but I can't agree with your methods of direct action"; who paternalistically feels that he can set the timetable for another man's freedom; who lives by the myth of time and who constantly advised the Negro to wait until a "more convenient season." Shallow understanding from people of good will is more frustrating than absolute misunderstanding from people of ill will. Lukewarm acceptance is much more bewildering than outright rejection.

I had hoped that the white moderate would understand that law and order exist for the purpose of establishing justice, and that when they fail to do this they become dangerously structured dams that block the flow of social progress. I had hoped that the white moderate would understand that the present tension of the South is merely a necessary phase of the transition from an obnoxious negative peace, where the Negro passively accepted his unjust plight, to a substance-filled positive peace, where all men will respect the dignity and worth of human personality. Actually, we who engage in nonviolent direct action are not the creators of tension. We merely bring to the surface the hidden tension that is already alive. We bring it out in the open where it can be seen and dealt with. Like a boil that can never be cured as long as it is covered up but must be opened with all its pus-flowing ugliness to the natural medicines of air and light, injustice must likewise be exposed, with all of the tension its exposing creates, to the light of human conscience and the air of national opinion before it can be cured.

In your statement you asserted that our actions, even though peaceful, must be condemned because they precipitate violence. But can this assertion be logically made? Isn't this like condemning the robbed man because his possession of money precipitated the evil act of robbery? Isn't this like condemning Socrates because his unswerving commitment to truth and his philosophical delvings precipitated the misguided popular mind to make him drink the hemlock? Isn't this like condemning Jesus because His unique God-consciousness and never-ceasing devotion to his will precipitated the evil act of crucifixion? We must come to see, as federal courts have consistently affirmed, that it is immoral to urge an individual to withdraw his efforts to gain his basic constitutional rights because the quest precipitates violence. Society must protect the robbed and punish the robber.

I had also hoped that the white moderate would reject the myth of time. I received a letter this morning from a white brother in Texas which said: "All Christians know that the colored people will receive equal rights eventually, but it is possible that you are in too great of a religious hurry. It has taken Christianity almost two thousand years to accomplish what it has. The teachings of Christ take time to come to earth." All that is said here grows out of a tragic misconception of time. It is the strangely irrational notion that there is something in the very flow of time that will inevitably cure all ills. Actually time is neutral. It can be used either destructively or constructively. I am coming to feel that the people of ill will have used time much more effectively than the people of good will. We will have to repent in this generation not merely for the vitriolic words and actions of the bad people, but for the appalling silence of the good people. We must come to see that human progress never rolls in on wheels of inevitability. It comes through the tireless efforts and persistent work of men willing to be co-workers with God, and without this hard work time itself becomes an ally of the forces of social stagnation. We must use time creatively, and forever realize that the time is always ripe to do right. Now is the time to make real the promise of democracy, and transform our pending national elegy into a creative psalm of brotherhood. Now is the time to lift our national policy from the quicksand of racial injustice to the solid rock of human dignity.

You spoke of our activity in Birmingham as extreme. At first I was rather disappointed that fellow clergymen would see my nonviolent efforts as those of the extremist. I started thinking about the fact that I stand in the middle of two opposing forces in the Negro community. One is a force of complacency made up of Negroes who, as a result of long years of oppression, have been so completely drained of self-respect and a sense of "somebodiness" that they have adjusted to segregation, and, of a few Negroes in the middle class who, because of a degree of academic and economic security, and because at points they profit by segregation, have unconsciously become insensitive to the problems of the masses. The other force is one of bitterness and hatred, and comes perilously close to advocating violence. It is expressed in the various black nationalist groups that are springing up over the nation, the largest and best known being Elijah Muhammad's Muslim movement. This movement is nourished by the contemporary frustration over the continued existence of racial discrimination. It is made up of people who have lost faith in America, who have absolutely repudiated Christianity, and who have concluded that the white man is an incurable "devil." I have tried to stand between these two forces, saying that we need not follow the "do-nothingism" of the complacent or the hatred and despair of the black nationalist. There is the more excellent way of love and nonviolent protest. I'm grateful to God that, through the Negro church, the dimension of nonviolence entered our struggle. If this philosophy had not emerged, I am convinced that by now many streets of the South would be flowing with floods of blood. And I am further convinced that if our white brothers dismiss as "rabble-rousers" and "outside agitators" those of us who are working through the channels of nonviolent direct action and refuse to support our nonviolent efforts, millions of Negroes, out of frustration and despair, will seek solace and security in black nationalist ideologies, a development that will lead inevitably to a frightening racial nightmare.

Oppressed people cannot remain oppressed forever. The urge for freedom will eventually come. This is what happened to the American Negro. Something within has reminded him of his birthright of freedom; something without has reminded him that he can gain it. Consciously and unconsciously, he has been swept in by what the Germans call the *Zeitgeist,* and with his black brothers of Africa, and his brown and yellow brothers of Asia, South America and the Caribbean, he is moving with a sense of cosmic urgency toward the promised land of racial justice. Recognizing this vital urge that has engulfed the Negro community, one should readily understand public demonstrations. The Negro has many pent-up resentments and latent frustrations. He has to get them out. So let him march sometime; let him have his prayer pilgrimages to the city hall; understand why he must have sit-ins and freedom rides. If his repressed emotions do not come out in these nonviolent ways, they will come out in ominous expressions of violence. This is not a threat; it is a fact of history. So I have not said to my people "get rid of your discontent." But I have tried to say that this normal and healthy discontent can be channelized through the creative outlet of nonviolent direct action. Now this approach is being dismissed as extremist. I must admit that I was initially disappointed in being so categorized.

But as I continued to think about the matter I gradually gained a bit of satisfaction from being considered an extremist. Was not Jesus an extremist in love—"Love your enemies, bless them that curse you, pray for them that despitefully use you." Was not Amos an extremist for justice—"Let justice roll down like waters and righteousness like a mighty stream." Was not Paul an extremist for the gospel of Jesus Christ—"I bear in my body the marks of the Lord Jesus." Was not Martin Luther an extremist—"Here I stand; I can do none other so help me God." Was not John Bunyan an extremist—"I will stay in jail to the end of my days before I make a butchery of my conscience." Was not Abraham Lincoln an extremist—"This nation cannot survive half slave and half free." Was not Thomas Jefferson an extremist—"We hold these truths to be self-evident, that all men are created equal." So the question is not whether we will be extremist but what kind of extremist will we be. Will we be extremists for hate or will we be extremists for love? Will we be extremists for the preservation of injustice—or will we be extremists for the cause of justice? In that dramatic scene on Calvary's hill, three men were crucified. We must not forget that all three were crucified for the same crime—the crime of extremism. Two were extremists for immorality, and thusly fell below their environment. The other, Jesus Christ, was an extremist for love, truth and goodness, and thereby rose above his environment. So, after all, maybe the South, the nation and the world are in dire need of creative extremists.

I had hoped that the white moderate would see this. Maybe I was too optimistic. Maybe I expected too much. I guess I should have realized that few members of a race that has oppressed another race can understand or appreciate the deep groans and passionate yearnings of those that have been oppressed and still fewer have the vision to see that injustice must be rooted out by strong, persistent and determined action. I am thankful, however, that some of our white brothers have grasped the meaning of this social revolution and committed themselves to it. They are still all too small in quantity, but they are big in quality. Some like Ralph McGill, Lillian Smith, Harry Golden and James Dabbs have written about our struggle in eloquent, prophetic and understanding terms. Others have marched with us down nameless streets of the South. They have languished in filthy roach-infested jails, suffering the abuse and brutality of angry policemen who see them as "dirty nigger-lovers." They, unlike so many of their moderate brothers and sisters, have recognized the urgency of the moment and sensed the need for powerful "action" antidotes to combat the disease of segregation.

Let me rush on to mention my other disappointment. I have been so greatly disappointed with the white church and its leadership. Of course, there are some notable exceptions. I am not unmindful of the fact that each of you has taken some significant stands on this issue. I commend you, Rev. Stallings, for your Christian stance on this past Sunday, in welcoming Negroes to your worship service on a non-segregated basis. I commend the Catholic leaders of this state for integrating Springhill College several years ago.

But despite these notable exceptions I must honestly reiterate that I have been disappointed with the church. I do not say that as one of the negative critics who can always find something wrong with the church. I say it as a minister of the gospel, who loves the church; who was nurtured in its bosom; who has been sustained by its spiritual blessings and who will remain true to it as long as the cord of life shall lengthen.

I had the strange feeling when I was suddenly catapulted into the leadership of the bus protest in Montgomery several years ago that we would have the support of the white church. I felt that the white ministers, priests and rabbis of the South would be some of our strongest allies. Instead, some have been outright opponents, refusing to understand the freedom movement and misrepresenting its leaders; all too many others have been more cautious than courageous and have remained silent behind the anesthetizing security of the stained-glass windows.

In spite of my shattered dreams of the past, I came to Birmingham with the hope that the white religious leadership of this community would see the justice of our cause, and with deep moral concern, serve as the channel through which our just grievances would get to the power structure. I had hoped that each of you would understand. But again I have been disappointed. I have heard numerous religious leaders of the South call upon their worshippers to comply with a desegregation decision because it is the *law,* but I have longed to hear white ministers say, "Follow this decree because integration is morally *right* and the Negro is your brother." In the midst of blatant injustices inflicted upon the Negro, I have watched white churches stand on the sideline and merely mouth pious irrelevancies and sanctimonious trivialities. In the midst of a mighty struggle to rid our nation of racial and economic injustice, I have heard so many ministers say, "Those are social issues with which the gospel has no real concern," and I have watched so many churches commit themselves to a completely otherworldly religion which made a strange distinction between body and soul, the sacred and the secular.

So here we are moving toward the exit of the twentieth century with a religious community largely adjusted to the status quo, standing as a taillight behind other community agencies rather than a headlight leading men to higher levels of justice.

I have traveled the length and breadth of Alabama, Mississippi and all the other southern states. On sweltering summer days and crisp autumn mornings I have looked at her beautiful churches with their lofty spires pointing heavenward. I have beheld the impressive outlay of her massive religious education buildings. Over and over again I have found myself asking: "What kind of people worship here? Who is their God? Where were their voices when the lips of Governor Barnett dripped with words of interposition and nullification? Where were they when Governor Wallace gave the clarion call for defiance and hatred? Where were their voices of support when tired, bruised and weary Negro men and women decided to rise from the dark dungeons of complacency to the bright hills of creative protest?"

Yes, these questions are still in my mind. In deep disappointment, I have wept over the laxity of the church. But be assured that my tears have been tears of love. There can be no deep disappointment where there is not deep love. Yes, I love the church; I love her sacred walls. How could I do otherwise? I am in the rather unique position of being the son, the grandson and the great-grandson of preachers. Yes, I see the church as the body of Christ. But, oh! How we have blemished and scarred that body through social neglect and fear of being nonconformists.

There was a time when the church was very powerful. It was during that period when the early Christians rejoiced when they were deemed worthy to suffer for what they believed. In those days the church was not merely a thermometer that recorded the ideas and principles of popular opinion; it was a thermostat that transformed the mores of society. Wherever the early Christians entered a town the power structure got disturbed and immediately sought to convict them for being "disturbers of the peace" and "outside agitators." But they went on with the conviction that they were "a colony of heaven," and had to obey God rather than man. They were small in number but big in commitment. They were too God-intoxicated to be "astronomically intimidated." They brought an end to such ancient evils as infanticide and gladiatorial contest.

Things are different now. The contemporary church is often a weak, ineffectual voice with an uncertain sound. It is so often the arch-supporter of the status quo. Far from being disturbed by the presence of the church, the power structure of the average community is consoled by the church's silent and often vocal sanction of things as they are.

But the judgment of God is upon the church as never before. If the church of today does not recapture the sacrificial spirit of the early church, it will lose its authentic ring, forfeit the loyalty of millions, and be dismissed as an irrelevant social club with no meaning for the twentieth century. I am meeting young people every day whose disappointment with the church has risen to outright disgust.

Maybe again, I have been too optimistic. Is organized religion too inextricably bound to the status quo to save our nation and the world? Maybe I must turn my faith to the inner spiritual church, the church within the church, as the true *ecclesia* and the hope of the world. But again I am thankful to God that some noble souls from the ranks of organized religion have broken loose from the paralyzing chains of conformity and joined us as active partners in the struggle for freedom. They have left their secure congregations and walked the streets of Albany, Georgia, with us. They have gone through the highways of the South on tortuous rides for freedom. Yes, they have gone to jail with us. Some have been kicked out of their churches, and lost support of their bishops and fellow ministers. But they have gone with the faith that right defeated is stronger than evil triumphant. These men have been the leaven in the lump of the race. Their witness has been the spiritual salt that has preserved the true meaning of the gospel in these troubled times. They have carved a tunnel of hope through the dark mountain of disappointment.

I hope the church as a whole will meet the challenge of this decisive hour. But even if the church does not come to the aid of justice, I have no despair about the future. I have no fear about the outcome of our struggle in Birmingham, even if our motives are presently misunderstood. We will reach the goal of freedom in Birmingham and all over the nation, because the goal of America is freedom. Abused and scorned though we may be, our destiny is tied up with the destiny of America. Before the Pilgrims landed at Plymouth we were here. Before the pen of Jefferson etched across the pages of history the majestic words of the Declaration of Independence, we were here. For more than two centuries our foreparents labored in this country without wages; they made cotton king; and they built the homes of their masters in the midst of brutal injustice and shameful humiliation—and yet out of a bottomless vitality they continued to thrive and develop. If the inexpressible cruelties of slavery could not stop us, the opposition we now face will surely fail. We will win our freedom because the sacred heritage of our nation and the eternal will of God are embodied in our echoing demands.

I must close now. But before closing I am impelled to mention one other point in your statement that troubled me profoundly. You warmly commended the Birmingham police force for keeping "order" and "preventing violence." I don't believe you would have so warmly commended the police force if you had seen its angry violent

(Top) Martin Luther King, Sr. and Jr., 1950s; *(bottom)* Martin Luther King, Jr., being arrested, Montgomery, Alabama, 1958 *(photos by Charles Moore/ Black Star).*

March on Washington, August 28, 1963. *(Above)* The crowd as viewed from the Lincoln Memorial *(photo by Fred Ward/Black Star)*; *(opposite page, top)* King giving "I Have a Dream" speech at the Lincoln Memorial *(Flip Schulke/Black Star)*; *(opposite page, bottom)* Demonstrators near the White House *(Fred Ward/Black Star)*.

WE
MARCH
FOR
EFFECTIVE
CIVIL RIGHTS
LAWS
NOW!

(Opposite page, top) King at Ebenezer Baptist Church, Atlanta, November 1964 *(photo by Flip Schulke/Black Star)*; *(opposite page, bottom)* from left, Mr. and Mrs. Martin Luther King, Sr., Coretta Scott King, Martin Luther King, Jr., and their four children, 1964 *(Scheler/Black Star)*; *(this page; top)* King and his wife Coretta in Washington, D.C. after returning from Scandinavia where he had gone to receive the Nobel Peace Prize, 1964 *(Fred Ward/Black Star)*; *(this page, bottom)* King at demonstration in Boston, Massachusetts, April 1965 *(Ivan Massar/Black Star)*.

Selma to Montgomery march for voting rights, March 1965.

(Opposite page, top) from left, Ralph Abernathy, King, Charles Evers, Ralph Bunche, and Rabbi Abraham Heschel at the start of the march *(photo by Matt Herron/Black Star)*; *(opposite page, bottom left)* King with two children during the march *(Vernon Merritt/Black Star)*; *(opposite page, bottom right)* giving "Our God Is Marching On!" address in front of the Alabama Statehouse in Montgomery *(Matt Herron/Black Star)*; *(this page, top)* on the march *(Vernon Merritt/Black Star)*; *(this page, bottom)* Martin and Coretta King leading the march *(Ivan Massar/ Black Star)*.

(Above) Martin Luther King, Jr. (1929–1968) *(photo by Flip Schulke/ Black Star)*; *(below)* Mule-drawn farm wagon carrying King's casket, April 1968 *(Black Star)*.

dogs literally biting six unarmed, nonviolent Negroes. I don't believe you would so quickly commend the policemen if you would observe their ugly and inhuman treatment of Negroes here in the city jail; if you would watch them push and curse old Negro women and young Negro girls; if you would see them slap and kick old Negro men and young boys; if you will observe them, as they did on two occasions, refuse to give us food because we wanted to sing our grace together. I'm sorry that I can't join you in your praise for the police department.

It is true that they have been rather disciplined in their public handling of the demonstrators. In this sense they have been rather publicly "nonviolent." But for what purpose? To preserve the evil system of segregation. Over the last few years I have consistently preached that nonviolence demands that the means we use must be as pure as the ends we seek. So I have tried to make it clear that it is wrong to use immoral means to attain moral ends. But now I must affirm that it is just as wrong, or even more so, to use moral means to preserve immoral ends. Maybe Mr. Connor and his policemen have been rather publicly nonviolent, as Chief Pritchett was in Albany, Georgia, but they have used the moral means of nonviolence to maintain the immoral end of flagrant racial injustice. T. S. Eliot has said that there is no greater treason than to do the right deed for the wrong reason.

I wish you had commended the Negro sit-inners and demonstrators of Birmingham for their sublime courage, their willingness to suffer and their amazing discipline in the midst of the most inhuman provocation. One day the South will recognize its real heroes. They will be the James Merediths, courageously and with a majestic sense of purpose facing jeering and hostile mobs and the agonizing loneliness that characterizes the life of the pioneer. They will be old, oppressed, battered Negro women, symbolized in a seventy-two-year-old woman of Montgomery, Alabama, who rose up with a sense of dignity and with her people decided not to ride the segregated buses, and responded to one who inquired about her tiredness with ungrammatical profundity: "My feet is tired, but my soul is rested." They will be the young high school and college students, young ministers of the gospel and a host of their elders courageously and nonviolently sitting-in at lunch counters and willingly going to jail for conscience's sake. One day the South will know that when these disinherited children of God sat down at lunch counters they were in reality standing up for the best in the American dream and the most sacred values in our Judeo-Christian heritage, and thusly, carrying our whole nation back to those great wells of democracy which were dug deep by the Founding Fathers in the formulation of the Constitution and the Declaration of Independence.

Never before have I written a letter this long (or should I say a book?). I'm afraid that it is much too long to take your precious time. I can assure you that it would have been much shorter if I had been writing from a comfortable desk, but what else is there to do when you are alone for days in the dull monotony of a narrow jail cell other than write long letters, think strange thoughts, and pray long prayers?

If I have said anything in this letter that is an overstatement of the truth and is indicative of an unreasonable impatience, I beg you to forgive me. If I have said anything in this letter that is an understatement of the truth and is indicative of my having a patience that makes me patient with anything less than brotherhood, I beg God to forgive me.

I hope this letter finds you strong in the faith. I also hope that circumstances will soon make it possible for me to meet each of you, not as an integrationist or a civil rights leader, but as a fellow clergyman and a Christian brother. Let us all hope that the dark clouds of racial prejudice will soon pass away and the deep fog of misunderstanding will be lifted from our fear-drenched communities and in some not too distant tomorrow the radiant stars of love and brotherhood will shine over our great nation with all of their scintillating beauty.

Yours for the cause of Peace and Brotherhood,

MARTIN LUTHER KING, JR.

Martin Luther King, Jr., *Why We Can't Wait* (New York: Harper & Row, 1963, 1964). The American Friends Committee first published this essay as a pamphlet. It has probably been reprinted more than anything else Dr. King wrote.

11

I Have a Dream

(1963)

The year 1963 was the centennial of the signing of the Emancipation Proclamation. It was truly a momentous year in American history and in the life of Martin Luther King, Jr.

Despite opposition from the governors of Alabama and Mississippi, the president of the United States, John F. Kennedy, authorized federal marshals to escort a few black students to register at the University of Mississippi and the University of Alabama. "Bull" Connor, the head of Birmingham, Alabama's, police department ordered his officers to turn fire hoses and police dogs on young demonstrators; as television cameras captured this horrible scene, the nation gasped in disbelief and revulsion. Medgar Evers, a thirty-seven-year-old NAACP field secretary in Jackson, Mississippi, was murdered on his front porch on June 12. Riots occurred throughout the summer. The nation stood on the brink of racial civil war. It needed a prophet who could help see through the smoke left by gunpowder and bombs.

Martin Luther King, Jr., who published *Why We Can't Wait* at this time, was the prophet of the hour. Although many of the phrases and themes that appear in "I Have a Dream" had often been repeated by Dr. King, this is his most well-known and most often quoted speech. He delivered it before the Lincoln Memorial on August 28, 1963, as the keynote address of the March on Washington, D.C., for Civil Rights. Television cameras allowed the entire nation to hear and see him plead for justice and freedom. Mrs. Coretta King once commented, "At that moment it seemed as if the Kingdom of God appeared. But it only lasted for a moment."

I AM HAPPY TO JOIN WITH YOU TODAY IN WHAT WILL GO DOWN IN HISTORY as the greatest demonstration for freedom in the history of our nation.

Fivescore years ago, a great American, in whose symbolic shadow we stand today, signed the Emancipation Proclamation. This momentous decree came as a great beacon light of hope to millions of Negro slaves who had been seared in the flames of withering injustice. It came as a joyous daybreak to end the long night of their captivity.

But one hundred years later, the Negro still is not free; one hundred years later, the life of the Negro is still sadly crippled by the manacles of segregation and the chains of discrimination; one hundred years later, the Negro lives on a lonely island of poverty in the midst of a vast ocean of material prosperity; one hundred years later, the Negro is still languished in the corners of American society and finds himself in exile in his own land.

So we've come here today to dramatize a shameful condition. In a sense we've come to our nation's capital to cash a check. When the architects of our republic wrote the magnificent words of the Constitution and the Declaration of Independence, they were signing a promissory note to which every American was to fall heir. This note was the promise that all men, yes, black men as well as white men, would be guaranteed the unalienable rights of life, liberty, and the pursuit of happiness.

It is obvious today that America has defaulted on this promissory note in so far as her citizens of color are concerned. Instead of honoring this sacred obligation, America has given the Negro people a bad check; a check which has come back marked "insufficient funds." We refuse to believe that there are insufficient funds in the great vaults of opportunity of this nation. And so we've come to cash this check, a check that will give us upon demand the riches of freedom and the security of justice.

We have also come to this hallowed spot to remind America of the fierce urgency of now. This is no time to engage in the luxury of cooling off or to take the tranquilizing drug of gradualism. Now is the time to make real the promises of democracy; now is the time to rise from the dark and desolate valley of segregation to the sunlit path of racial justice; now is the time to lift our nation from the quicksands of racial injustice to the solid rock of brotherhood; now is the time to make justice a reality for all God's children. It would be fatal for the nation to overlook the urgency of the moment. This sweltering summer of the Negro's legitimate discontent will not pass until there is an invigorating autumn of freedom and equality.

Nineteen sixty-three is not an end, but a beginning. And those who hope that the Negro needed to blow off steam and will now be content, will have a rude awakening if the nation returns to business as usual.

There will be neither rest nor tranquility in America until the Negro is granted his citizenship rights. The whirlwinds of revolt will continue to shake the foundations of our nation until the bright day of justice emerges.

But there is something that I must say to my people who stand on the warm threshold which leads into the palace of justice. In the process of gaining our rightful place we must not be guilty of wrongful deeds.

Let us not seek to satisfy our thirst for freedom by drinking from the cup of bitterness and hatred. We must forever conduct our struggle on the high plane of dignity and discipline. We must not allow our creative protest to degenerate into physical violence. Again and again we must rise to the majestic heights of meeting physical force with soul force.

The marvelous new militancy which has engulfed the Negro community must not lead us to a distrust of all white people, for many of our white brothers, as evidenced by their presence here today, have come to realize that their destiny is tied up with our destiny and they have come to realize that their freedom is inextricably bound to our freedom. This offense we share mounted to storm the battlements of injustice must be carried forth by a biracial army. We cannot walk alone.

And as we walk, we must make the pledge that we shall always march ahead. We cannot turn back. There are those who are asking the devotees of civil rights, "When will you be satisfied?" We can never be satisfied as long as the Negro is the victim of the unspeakable horrors of police brutality.

We can never be satisfied as long as our bodies, heavy with fatigue of travel, cannot gain lodging in the motels of the highways and the hotels of the cities. We cannot be satisfied as long as the Negro's basic mobility is from a smaller ghetto to a larger one.

We can never be satisfied as long as our children are stripped of their selfhood and robbed of their dignity by signs stating "for whites only." We cannot be satisfied as long as a Negro in Mississippi cannot vote and a Negro in New York believes he has nothing for which to vote. No, we are not satisfied, and we will not be satisfied until justice rolls down like waters and righteousness like a mighty stream.

I am not unmindful that some of you come here out of excessive trials and tribulation. Some of you have come fresh from narrow jail cells. Some of you have come from areas where your quest for freedom left you battered by the storms of persecution and staggered by the winds of police brutality. You have been the veterans of creative suffering. Continue to work with the faith that unearned suffering is redemptive.

Go back to Mississippi; go back to Alabama; go back to South Carolina; go back to Georgia; go back to Louisiana; go back to the slums and ghettos of the northern cities, knowing that somehow this situation can, and will be changed. Let us not wallow in the valley of despair.

So I say to you, my friends, that even though we must face the difficulties of today and tomorrow, I still have a dream. It is a dream deeply rooted in the American dream that one day this nation will rise up and live out the true meaning of its creed—we hold these truths to be self-evident, that all men are created equal.

I have a dream that one day on the red hills of Georgia, sons of former slaves and sons of former slave-owners will be able to sit down together at the table of brotherhood.

I have a dream that one day, even the state of Mississippi, a state sweltering with the heat of injustice, sweltering with the heat of oppression, will be transformed into an oasis of freedom and justice.

I have a dream my four little children will one day live in a nation where they will not be judged by the color of their skin but by the content of their character. I have a dream today!

I have a dream that one day, down in Alabama, with its vicious racists, with its governor having his lips dripping with the words of interposition and nullification, that one day, right there in Alabama, little black boys and black girls will be able to join hands with little white boys and white girls as sisters and brothers. I have a dream today!

I have a dream that one day every valley shall be exalted, every hill and mountain shall be made low, the rough places shall be made plain, and the crooked places shall be made straight and the glory of the Lord will be revealed and all flesh shall see it together.

This is our hope. This is the faith that I go back to the South with.

With this faith we will be able to hew out of the mountain of despair a stone of hope. With this faith we will be able to transform the jangling discords of our nation into a beautiful symphony of brotherhood.

With this faith we will be able to work together, to pray together, to struggle together, to go to jail together, to stand up for freedom together, knowing that we will be free one day. This will be the day when all of God's children will be able to sing with new meaning—"my country 'tis of thee; sweet land of liberty; of thee I sing; land where my fathers died, land of the pilgrim's pride; from every mountain side, let freedom ring"—and if America is to be a great nation, this must become true.

So let freedom ring from the prodigious hilltops of New Hampshire.

Let freedom ring from the mighty mountains of New York.

Let freedom ring from the heightening Alleghenies of Pennsylvania.

Let freedom ring from the snow-capped Rockies of Colorado.

Let freedom ring from the curvaceous slopes of California.

But not only that.

Let freedom ring from Stone Mountain of Georgia.

Let freedom ring from Lookout Mountain of Tennessee.

Let freedom ring from every hill and molehill of Mississippi, from every mountainside, let freedom ring.

And when we allow freedom to ring, when we let it ring from every village and hamlet, from every state and city, we will be able to speed up that day when all of God's children—black men and white men, Jews and Gentiles, Catholics and Protestants—will be able to join hands and to sing in the words of the old Negro spiritual, "Free at last, free at last; thank God Almighty, we are free at last."

Negro History Bulletin 21 (May 1968): 16–17.

12

Nobel Prize Acceptance Speech

(1964)

On November 22, 1963, President John Fitzgerald Kennedy was assassinated. Dr. King and Malcolm X, the great shaman of the Nation of Islam, both declared that America was a sick society. They both saw that the pathologies of racial hatred and injustice that had been tolerated against people of African descent had almost caused a fatal coronary within the body politic. Even if segregationists within the city halls, state legislatures, governor's mansions, and the United States Congress itself could not see this, the world paused for a moment to analyze what had happened. Vice President Lyndon Johnson was immediately sworn in to the presidency. To his credit, and with the help of an armada of secular and religious activists, he used his immense legislative genius and influence to defeat congressional segregationists who had been opposing President Kennedy's Civil Rights Bill. By a vote of 71 to 29, the United States Senate imposed cloture to end a southern filibuster on June 10, 1964. By July 2 the landmark Civil Rights Bill of 1964 was enacted and signed by President Johnson.

For many black urban dwellers, who languished in poverty and joblessness, this seemed like too little too late. Riots erupted in New York state. They began in Harlem and Brooklyn on July 18. When they occurred in Rochester, New York, on July 25, Governor Nelson Rockefeller ordered the national guard to restore order to that city. On August 2, Jersey City, New Jersey, followed suit.

Two days later, the dead and mutilated bodies of three civil rights workers who had been missing since June 21 were discovered on a farm near Philadelphia, Mississippi. The world watched the most powerful nation on the planet plunge deeper into a prolonged period of African American rebellion against racial injustice. As the Nobel Prize Committee scanned the globe to designate the greatest apostle of peace, they chose the Reverend Dr. Martin Luther King, Jr. The speech that follows is the full text of Dr. King's acceptance address after receiving the Nobel Peace Prize in Oslo, Norway, on December 10, 1964.

When once asked by an interviewer what the significance of being the recipient of this much coveted award was, Dr. King replied, "The Nobel award recognizes the amazing discipline of the Negro. Though we have had riots, the bloodshed we would have known without the discipline of nonviolence would have been frightening."

YOUR MAJESTY, YOUR ROYAL HIGHNESS, MR. PRESIDENT, EXCELLENCIES, ladies and gentlemen:

I accept the Nobel Prize for Peace at a moment when twenty-two million Negroes of the United States of America are engaged in a creative battle to end the long night of racial injustice. I accept this award in behalf of a civil rights movement which is moving with determination and a majestic scorn for risk and danger to establish a reign of freedom and rule of justice.

I am mindful that only yesterday in Birmingham, Alabama, our children, crying out for brotherhood, were answered with fire hoses, snarling dogs and even death. I am mindful that only yesterday in Philadelphia, Mississippi, young people seeking to secure the right to vote were brutalized and murdered.

I am mindful that debilitating and grinding poverty afflicts my people and chains them to the lowest rung of the economic ladder.

Therefore, I must ask why this prize is awarded to a movement which is beleaguered and committed to unrelenting struggle: to a movement which has not won the very peace and brotherhood which is the essence of the Nobel Prize.

After contemplation, I conclude that this award which I receive on behalf of that movement is profound recognition that nonviolence is the answer to the crucial political and moral question of our time—the need for man to overcome oppression and violence without resorting to violence and oppression.

Civilization and violence are antithetical concepts. Negroes of the United States, following the people of India, have demonstrated that nonviolence is not servile passivity, but a powerful moral force which makes for social transformation. Sooner or later, all people of the world will have to discover a way to live together in peace, and thereby transform this pending cosmic elegy into a creative psalm of brotherhood.

If this is to be achieved, man must evolve for all human conflict a method which rejects revenge, aggression and retaliation. The foundation of such a method is love.

From the depths of my heart I am aware that this prize is much more than an honor to me personally.

Every time I take a flight I am always mindful of the many people who make a successful journey possible, the known pilots and the unknown ground crew.

So you honor the dedicated pilots of our struggle who have sat at the controls as the freedom movement soared into orbit. You honor, once again, Chief (Albert) Luthuli of South Africa, whose struggles with and for his people, are still met with the most brutal expression of man's inhumanity to man.

You honor the ground crew without whose labor and sacrifices the jet flights to freedom could never have left the earth.

Most of these people will never make the headlines and their names will not appear in *Who's Who*. Yet the years have rolled past and when the blazing light of truth is focused on this marvelous age in which we live—men and women will know and children will be taught that we have a finer land, a better people, a more noble civilization—because these humble children of God were willing to suffer for righteousness' sake.

I think Alfred Nobel would know what I mean when I say that I accept this award in the spirit of the curator of some precious heirloom which he holds in trust for its true owners—all those to whom beauty is truth and truth beauty—and in whose eyes the beauty of genuine brotherhood and peace is more precious than diamonds or silver or gold.

The tortuous road which has led from Montgomery, Alabama, to Oslo bears witness to this truth. This is a road over which millions of Negroes are travelling to find a new sense of dignity. This same road has opened for all Americans a new era of progress and hope. It has led to a new civil rights bill, and it will, I am convinced, be widened and lengthened into a superhighway of justice as Negro and white men in increasing numbers create alliances to overcome their common problems.

I accept this award today with an abiding faith in America and an audacious faith in the future of mankind. I refuse to accept the idea that the "isness" of man's present nature makes him morally incapable of reaching up for the eternal "oughtness" that forever confronts him.

I refuse to accept the idea that man is mere flotsam and jetsam in the river of life which surrounds him. I refuse to accept the view that mankind is so tragically bound to the starless midnight of racism and war that the bright daybreak of peace and brotherhood can never become a reality.

I refuse to accept the cynical notion that nation after nation must spiral down a militaristic stairway into a hell of thermonuclear destruction. I believe that unarmed truth and unconditional love will have the final word in reality. That is why right temporarily defeated is stronger than evil triumphant.

I believe that even amid today's mortar bursts and whining bullets, there is still hope for a brighter tomorrow. I believe that wounded justice, lying prostrate on the blood-flowing streets of our nations, can be lifted from this dust of shame to reign supreme among the children of men.

I have the audacity to believe that peoples everywhere can have three meals a day for their bodies, education and culture for their minds, and dignity, equality and freedom for their spirits. I believe that what self-centered men have torn down men other-centered can build up. I still believe that one day mankind will bow before the altars of God and be crowned triumphant over war and bloodshed, and nonviolent redemptive good will proclaim the rule of the land. "And the lion and the lamb shall lie down together and every man shall sit under his own vine and fig tree and none shall be afraid." I still believe that we shall overcome.

This faith can give us courage to face the uncertainties of the future. It will give our tired feet new strength as we continue our forward stride toward the city of freedom. When our days become dreary with low-hovering clouds and our nights become darker than a thousand midnights, we will know that we are living in the creative turmoil of a genuine civilization struggling to be born.

Today I come to Oslo as a trustee, inspired and with renewed dedication to humanity. I accept this prize on behalf of all men who love peace and brotherhood.

Negro History Bulletin 21 (May 1968): 16–17.

PART III

The Dream Is Deferred

(1963–1968)

13

Eulogy for the Martyred Children

(1963)

Shortly after the euphoria of the successful March on Washington in August 1963, the nation was sickened again by the malice of white segregationists. As many progressive people around the world were praising the genius of the great African American scholar and activist Dr. William E. B. Du Bois (1868–1963), who died on August 27, bigots were plotting to disrupt and dismiss recent landmarks for building a more just society.

The Reverend Dr. King delivered this sermon at the funeral of the little girls who were killed on September 15, 1963, by a bomb as they attended the Sunday school of the 16th Street Baptist Church in Birmingham, Alabama.

THIS AFTERNOON WE GATHER IN THE QUIET OF THIS SANCTUARY TO PAY our last tribute of respect to these beautiful children of God. They entered the stage of history just a few years ago, and in the brief years that they were privileged to act on this mortal stage, they played their parts exceedingly well. Now the curtain falls; they move through the exit; the drama of their earthly life comes to a close. They are now committed back to that eternity from which they came.

These children—unoffending; innocent and beautiful—were the victims of one of the most vicious, heinous crimes ever perpetrated against humanity.

Yet they died nobly. They are the martyred heroines of a holy crusade for freedom and human dignity. So they have something to say to us in their death. They have something to say to every minister of the gospel who has remained silent behind the safe security of stained-glass windows. They have something to say to every politician who has fed his constituents the stale bread of hatred and the spoiled meat of racism. They have something to say to a federal government that has compromised with the undemocratic practices of southern dixiecrats and the blatant hypocrisy of right-wing northern Republicans. They have something to say to every Negro who passively accepts the evil system of segregation, and stands on the sidelines in the midst of a mighty struggle for justice. They say to each of us, black and white alike, that we must substitute courage for caution. They say to us that we must be concerned not merely about *WHO* murdered them, but about the system, the way of life and the philosophy which *PRODUCED* the murderers. Their death says to us that we must work passionately and unrelentingly to make the American dream a reality.

So they did not die in vain. God still has a way of wringing good out of evil. History has proven over and over again that unmerited suffering is redemptive. The innocent blood of these little girls may well serve as the redemptive force that will bring new light to this dark city. The holy Scripture says, "A little child shall lead them." The death of these little children may lead our whole Southland from the low road of man's inhumanity to man to the high road of peace and brotherhood. These tragic deaths may lead our nation to substitute an aristocracy of character for an aristocracy of color. The spilt blood of these innocent girls may cause the whole citizenry of Birmingham to transform the negative extremes of a dark past into the positive extremes of a bright future. Indeed, this tragic event may cause the white South to come to terms with its conscience.

So in spite of the darkness of this hour we must not despair. We must not become bitter; nor must we harbor the desire to retaliate with violence. We must not lose faith in our white brothers. Somehow we must believe that the most misguided among them can learn to respect the dignity and worth of all human personality.

May I now say a word to you, the members of the bereaved families. It is almost impossible to say anything that can console you at this difficult hour and remove the deep clouds of disappointment which are floating in your mental skies. But I hope you can find a little consolation from the universality of this experience. Death comes to every individual. There is an amazing democracy about death. It is not aristocracy for some of the people, but a democracy for all of the people. Kings die and beggars die; rich men die and poor men die; old people die and young people die; death comes to the innocent and it comes to the guilty. Death is the irreducible common denominator of all men.

I hope you can find some consolation from Christianity's affirmation that death is not the end. Death is not a period that ends the great sentence of life, but a comma that punctuates it to more lofty significance. Death is not a blind alley that leads the human race into a state of nothingness, but an open door which leads man into life eternal. Let this daring faith, this great invincible surmise, be your sustaining power during these trying days.

At times, life is hard, as hard as crucible steel. It has its bleak and painful moments. Like the ever-flowing waters of a river, life has its moments of drought and its moments of flood. Like the ever-changing cycle of the seasons, life has the soothing warmth of the summers and the piercing chill of its winters. But through it all, God walks with us. Never forget that God is able to lift you from fatigue of despair to the buoyancy of hope, and transform dark and desolate valleys into sunlit paths of inner peace.

Your children did not live long, but they lived well. The quantity of their lives was disturbingly small, but the quality of their lives was magnificently big. Where they died and what they were doing when death came will remain a marvelous tribute to each of you and an eternal epitaph to each of them. They died not in a den or dive nor were they hearing and telling filthy jokes at the time of their death. They died within the sacred walls of the church after discussing a principle as eternal as love.

Shakespeare had Horatio utter some beautiful words over the dead body of Hamlet. I paraphrase these words today as I stand over the last remains of these lovely girls.

"Good-night sweet princesses; may the flight of angels take thee to thy eternal rest."

Epilogue: The doors of the 16th Street Baptist Church reopened on Sunday, June 7, 1964.

The "reentry" sermon was preached by a white clergyman, the Reverend H. O. Hester, secretary of the Department of Missions, Alabama Baptist Convention.

Offprint in the Library of the Martin Luther King, Jr., Center for Nonviolent Social Change, Atlanta, Georgia.

14

Our God Is Marching On!

(1965)

Despite the passage of the Civil Rights Act of 1964, African Americans were still denied the privilege of voting in many areas of the South. They were often intimidated, and gerrymandering and other forms of corruption were used against them. During the summer 1964 debate over the nominations for the presidency of the United States, the Democratic National Convention had refused to seat the duly elected delegates from the Mississippi Freedom Party.

Politics as usual illustrated to many progressive leaders that far more needed to be done to address political and social equality. As a consequence, the Southern Christian Leadership Conference, Student Nonviolent Coordinating Committee, and other civil rights groups joined forces in Selma, Alabama, in order to "dramatize" black disfranchisement. At the height of the demonstrations in Selma between February 1 and 4, 1965, 3,000 people were arrested. Progress seemed to be more apparent in the courts than in the streets, however.

On February 4, a federal court banned literacy testing and other silly technicalities used to deny African Americans the right to vote. But the murder of Malcolm X in New York City on February 21 fanned the burning suspicion that government surveillance and sabotage was the nation's answer to racial injustice rather than justice itself. After three white segregationists murdered the Reverend James Reeb, a white Unitarian minister, on the streets of Selma on March 9, many more white people began to demand that the government address the issues of black disenfranchisement.

Quoting the great hymn of the Civil Rights Movement, President Lyndon Johnson declared before a joint session of Congress that "We shall overcome!" He pleaded with Congress to pass legislation that would guarantee voting rights, and condemned what he characterized as "the crippling legacy of bigotry and injustice." But more pressure was needed. Dr. King sent out a call for all people of good will to come to Selma so that the entire nation would be able to see the level of support and the extent of the bigotry. Later, after leading a 54-mile march from Selma to Montgomery, Dr. King spoke triumphantly before the state capitol building in Montgomery, often called "the Cradle of the Confederacy." He defended the march from Selma's Edmund Pettus Bridge, where marchers had been beaten and turned back by Alabama state troopers on their first attempts to walk to Montgomery. The march had finally begun in earnest on March 21, and it ended with this speech on March 25, 1965.

MY DEAR AND ABIDING FRIENDS, RALPH ABERNATHY, AND TO ALL THE distinguished Americans seated here on the rostrum, my friends and co-workers of the state of Alabama and to all of the freedom-loving people who have assembled here this afternoon, from all over our nation and from all over the world.

Last Sunday, more than eight thousand of us started on a mighty walk from Selma, Alabama. We have walked on meandering highways and rested our bodies on rocky byways. Some of our faces are burned from the outpourings of the sweltering sun. Some have literally slept in the mud. We have been drenched by the rains.

Our bodies are tired, and our feet are somewhat sore, but today as I stand before you and think back over that great march, I can say as Sister Pollard said, a seventy-year-old Negro woman who lived in this community during the bus boycott and one day she was asked while walking if she wanted a ride and when she answered, "No," the person said, "Well, aren't you tired?" And with her ungrammatical profundity, she said, "My feets is tired, but my soul is rested."

And in a real sense this afternoon, we can say that our feet are tired, but our souls are rested.

"WE ARE HERE"

They told us we wouldn't get here. And there were those who said that we would get here only over their dead bodies, but all the world today knows that we are here and that we are standing before the forces of power in the state of Alabama saying, "We ain't goin' let nobody turn us around."

The Civil Rights Act of 1964 gave Negroes some part of their rightful dignity, but without the vote it was dignity without strength.

Once more the method of nonviolent resistance was unsheathed from its scabbard and once again an entire community was mobilized to confront the adversary. And again the brutality of a dying order shrieks across the land. Yet Selma, Alabama, became a shining moment in the conscience of man.

There never was a moment in American history more honorable and more inspiring than the pilgrimage of clergymen and laymen of every race and faith pouring into Selma to face danger at the side of its embattled Negroes.

Confrontation of good and evil compressed in the tiny community of Selma generated the massive power to turn the whole nation to a new course. A president born in the South had the sensitivity to feel the will of the country, and in an address that will live in history as one of the most passionate pleas for human rights ever made by a president of our nation, he pledged the might of the federal government to cast off the centuries-old blight. President Johnson rightly praised the courage of the Negro for awakening the conscience of the nation.

On our part we must pay our profound respects to the white Americans who cherish their democratic traditions over the ugly customs and privileges of generations and come forth boldly to join hands with us. From Montgomery to Birmingham, from Birmingham to Selma, from Selma back to Montgomery, a trail wound in a circle and often bloody, yet it has become a highway up from darkness. Alabama has tried to nurture and defend evil, but the evil is choking to death in the dusty roads and streets of this state.

So I stand before you this afternoon with the conviction that segregation is on its deathbed in Alabama and the only thing uncertain about it is how costly the segregationists and Wallace will make the funeral.

Our whole campaign in Alabama has been centered around the right to vote. In focusing the attention of the nation and the world today on the flagrant denial of the right to vote, we are exposing the very origin, the root cause, of racial segregation in the Southland.

The threat of the free exercise of the ballot by the Negro and the white masses alike resulted in the establishing of a segregated society. They segregated southern money from the poor whites; they segregated southern mores from the rich whites; they segregated southern churches from Christianity; they segregated southern minds from honest thinking, and they segregated the Negro from everything.

We have come a long way since that travesty of justice was perpetrated upon the American mind. Today I want to tell the city of Selma, today I want to say to the state of Alabama, today I want to say to the people of America and the nations of the world: We are not about to turn around. We are on the move now. Yes, we are on the move and no wave of racism can stop us.

"WE ARE ON THE MOVE"

We are on the move now. The burning of our churches will not deter us. We are on the move now. The bombing of our homes will not dissuade us. We are on the move now. The beating and killing of our clergymen and young people will not divert us. We are on the move now. The arrest and release of known murderers will not discourage us. We are on the move now.

Like an idea whose time has come, not even the marching of mighty armies can halt us. We are moving to the land of freedom.

Let us therefore continue our triumph and march to the realization of the American dream. Let us march on segregated housing, until every ghetto of social and economic depression dissolves and Negroes and whites live side by side in decent, safe and sanitary housing.

Let us march on segregated schools until every vestige of segregated and inferior education becomes a thing of the past and Negroes and whites study side by side in the socially healing context of the classroom.

Let us march on poverty, until no American parent has to skip a meal so that their children may march on poverty, until no starved man walks the streets of our cities and towns in search of jobs that do not exist.

Let us march on ballot boxes, march on ballot boxes until race baiters disappear from the political arena. Let us march on ballot boxes until the Wallaces of our nation tremble away in silence.

Let us march on ballot boxes, until we send to our city councils, state legislatures, and the United States Congress men who will not fear to do justice, love mercy, and walk humbly with their God. Let us march on ballot boxes until all over Alabama God's children will be able to walk the earth in decency and honor.

For all of us today the battle is in our hands. The road ahead is not altogether a smooth one. There are no broad highways to lead us easily and inevitably to quick solutions. We must keep going.

"MY PEOPLE, LISTEN!"

My people, my people, listen! The battle is in our hands. The battle is in our hands in Mississippi and Alabama, and all over the United States.

So as we go away this afternoon, let us go away more than ever before committed to the struggle and committed to nonviolence. I must admit to you there are still some difficulties ahead. We are still in for a season of suffering in many of the black belt countries of Alabama, many areas of Mississippi, many areas of Louisiana.

I must admit to you there are still jail cells waiting for us, dark and difficult moments. We will go on with the faith that nonviolence and its power transformed dark yesterdays into bright tomorrows. We will be able to change all of these conditions.

Our aim must never be to defeat or humiliate the white man but to win his friendship and understanding. We must come to see that the end we seek is a society at peace with itself, a society that can live with its conscience. That will be a day not of the white man, not of the black man. That will be the day of man as man.

I know you are asking today, "How long will it take?" I come to say to you this afternoon however difficult the moment, however frustrating the hour, it will not be long, because truth pressed to earth will rise again.

How long? Not long, because no lie can live forever.

How long? Not long, because you still reap what you sow.

How long? Not long. Because the arm of the moral universe is long but it bends toward justice.

How long? Not long, 'cause mine eyes have seen the glory of the coming of the Lord, trampling out the vintage where the grapes of wrath are stored. He has loosed the fateful lightning of his terrible swift sword. His truth is marching on.

He has sounded forth the trumpets that shall never call retreat. He is lifting up the hearts of man before His judgment seat. Oh, be swift, my soul, to answer Him. Be jubilant, my feet. Our God is marching on.

Unpublished transcription of a recording of this speech provided by Richard Newman, noted historian of African American religion and an archivist at the New York Public Library.

15

Nonviolence: The Only Road to Freedom

(1966)

On May 4, 1966, well over eighty percent of registered African Americans in the state of Alabama voted in that state's Democratic Primary. This was a clear signal that the movement toward political equality was beginning to catch up with the quest for social equality. But there were black people who did not believe in the notion of integration. Many had become cynical about any possibility of acquiring racial justice from the white majority. They began to embrace the old idea of black self-determination that had been espoused by black nationalists such as Bishop Henry McNeal Turner (1834–1915) of the African Methodist Episcopal Church and later by Marcus Garvey (1887–1940) and his Universal Negro Improvement Association. The "election" of Stokely Carmichael to the presidency of Student Nonviolent Coordinating Committee on May 16, 1966, as well as his militant enunciation of the notion and slogan, "Black Power!", signalled the return of this older idea. Amid urban riots, a new wave of strident militancy within black America, stiff competition for funds and political influence from other civil rights organizations, and the signs that the nation's attention was increasingly being diverted from the Civil Rights Movement, Dr. King and his editorial staff defended the Southern Christian Leadership Conference's position in this *Ebony* magazine article. They argued that nonviolent resistance was the only effective strategy for social change available to black people.

THE YEAR 1966 BROUGHT WITH IT THE FIRST PUBLIC CHALLENGE TO THE philosophy and strategy of nonviolence from within the ranks of the civil rights movement. Resolutions of self-defense and Black Power sounded forth from our friends and brothers. At the same time riots erupted in several major cities. Inevitably a link was made between the two phenomena though movement leadership continued to deny any implications of violence in the concept of Black Power.

The nation's press heralded these incidents as an end of the Negro's reliance on nonviolence as a means of achieving freedom. Articles appeared on "The Plot to Get Whitey," and, "Must Negroes fight back?" and one had the impression that a serious movement was underway to lead the Negro to freedom through the use of violence.

Indeed, there was much talk of violence. It was the same talk we have heard on the fringes of the nonviolent movement for the past ten years. It was the talk of fearful men, saying that they would not join the nonviolent movement because they would not remain nonviolent if attacked. Now the climate had shifted so that it was even more popular to talk of violence, but in spite of the talk of violence there emerged no action in this direction. One reporter pointed out in a recent *New Yorker* article, that the fact that Beckwith, Price, Rainey, and Collie Leroy Wilkins remain alive is living testimony to the fact that the Negro remains nonviolent. And if this is not enough, a mere check of the statistics of casualties in the recent riots shows that the vast majority of persons killed in riots are Negroes. All the reports of sniping in Los Angeles's expressways did not produce a single casualty. The young demented white student at the University of Texas has shown what damage a sniper can do when he is serious. In fact, this one young man killed more people in one day than all the Negroes have killed in all the riots in all the cities since the Harlem riots of 1964. This must raise a serious question about the violent intent of the Negro, for certainly there are many ex-GIs within our ghettos, and no small percentage of those recent migrants from the South have demonstrated some proficiency hunting squirrels and rabbits.

I can only conclude that the Negro, even in his bitterest moments, is not intent on killing white men to be free. This does not mean that the Negro is a saint who abhors violence. Unfortunately, a check of the hospitals in any Negro community on any Saturday night will make you painfully aware of the violence within the Negro community. Hundreds of victims of shooting and cutting lie bleeding in the emergency rooms, but there is seldom if ever a white person who is the victim of Negro hostility.

I have talked with many persons in the ghettos of the North who argue eloquently for the use of violence. But I observed none of them in the mobs that rioted in Chicago. I have heard the street-corner preachers in Harlem and in Chicago's Washington Park, but in spite of the bitterness preached and the hatred espoused, none of them has ever been able to start a riot. So far, only the police through their fears and prejudice have goaded our people to riot. And once the riot starts, only the police or the National Guard have been able to put an end to them. This demonstrates that these violent eruptions are unplanned, uncontrollable temper tantrums brought on by long-neglected poverty, humiliation, oppression and exploitation. Violence as a strategy for social change in America is nonexistent. All the sound and fury seems but the posturing of cowards whose bold talk produces no action and signifies nothing.

I am convinced that for practical as well as moral reasons, nonviolence offers the only road to freedom for my people. In violent warfare, one must be prepared to face ruthlessly the fact that there will be casualties by the thousands. In Vietnam, the United States has evidently decided that it is willing to slaughter millions, sacrifice some two hundred thousand men and twenty billion dollars a year to secure the freedom of some fourteen million Vietnamese. This is to fight a war on Asian soil, where Asians are in the majority. Anyone leading a violent conflict must be willing to make a similar assessment regarding the possible casualties to a minority population confronting a well-armed, wealthy majority with a fanatical right wing that is capable of exterminating the entire black population and which would not hesitate such an attempt if the survival of white Western materialism were at stake.

Arguments that the American Negro is a part of a world which is two-thirds colored and that there will come a day when the oppressed people of color will rise together to throw off the yoke of white oppression are at least fifty years away from being relevant. There is no colored nation, including China, which now shows even the potential of leading a revolution of color in any international proportion. Ghana, Zambia, Tanzania and Nigeria are fighting their own battles for survival against poverty, illiteracy and the subversive influence of neocolonialism, so that they offer no hope to Angola, Southern Rhodesia and South Africa, and much less to the American Negro.

The hard cold facts of racial life in the world today indicate that the hope of the people of color in the world may well rest on the American Negro and his ability to reform the structures of racist imperialism from within and thereby turn the technology and wealth of the West to the task of liberating the world from want.

This is no time for romantic illusions about freedom and empty philosophical debate. This is a time for action. What is needed is a strategy for change, a tactical program which will bring the Negro into the mainstream of American life as quickly as possible. So far, this has only been offered by the nonviolent movement.

Our record of achievement through nonviolent action is already remarkable. The dramatic social changes which have been made across the South are unmatched in the annals of history. Montgomery, Albany, Birmingham and Selma have paved the way for untold progress. Even more remarkable is the fact that this progress occurred with a minimum of human sacrifice and loss of life.

Not a single person has been killed in a nonviolent demonstration. The bombings of the 16th Street Baptist Church occurred several months after demonstrations stopped. Rev. James Reeb, Mrs. Viola Liuzzo and Jimmie Lee Jackson were all murdered at night following demonstrations. And fewer people have been killed in ten years of action across the South than were killed in three nights of rioting in Watts. No similar changes have occurred without infinitely more sufferings, whether it be Gandhi's drive for independence in India or any African nation's struggle for independence.

THE QUESTION OF SELF-DEFENSE

There are many people who very honestly raise the question of self-defense. This must be placed in perspective. It goes without saying that people will protect their homes. This is a right guaranteed by the

Constitution and respected even in the worst areas of the South. But the mere protection of one's home and person against assault by lawless night riders does not provide any positive approach to the fears and conditions which produce violence. There must be some program for establishing law. Our experience in places like Savannah and Macon, Georgia, has been that a drive which registers Negroes to vote can do more to provide protection of the law and respect for Negroes by even racist sheriffs than anything we have seen.

In a nonviolent demonstration, self-defense must be approached from quite another perspective. One must remember that the cause of the demonstration is some exploitation or form of oppression that has made it necessary for men of courage and good will to demonstrate against the evil. For example, a demonstration against the evil of *de facto* school segregation is based on the awareness that a child's mind is crippled daily by inadequate educational opportunity. The demonstrator agrees that it is better for him to suffer publicly for a short time to end the crippling evil of school segregation than to have generation after generation of children suffer in ignorance.

In such a demonstration, the point is made that schools are inadequate. This is the evil to which one seeks to point; anything else detracts from that point and interferes with confrontation of the primary evil against which one demonstrates. Of course, no one wants to suffer and be hurt. But it is more important to get at the cause than to be safe. It is better to shed a little blood from a blow on the head or a rock thrown by an angry mob than to have children by the thousands grow up reading at a fifth- or sixth-grade level.

It is always amusing to me when a Negro man says that he can't demonstrate with us because if someone hit him he would fight back. Here is a man whose children are being plagued by rats and roaches, whose wife is robbed daily at overpriced ghetto food stores, who himself is working for about two-thirds the pay of a white person doing a similar job and with similar skills, and in spite of all this daily suffering it takes someone spitting on him or calling him a nigger to make him want to fight.

Conditions are such for Negroes in America that all Negroes ought to be fighting aggressively. It is as ridiculous for a Negro to raise the question of self-defense in relation to nonviolence as it is for a soldier on the battlefield to say he is not going to take any risks. He is there because he believes that the freedom of his country is worth the risk of his life. The same is true of the nonviolent demonstrator. He sees the misery of his people so clearly that he volunteers to suffer in their behalf and put an end to their plight.

Furthermore, it is extremely dangerous to organize a movement around self-defense. The line between defensive violence and aggressive or retaliatory violence is a fine line indeed. When violence is tolerated even as a means of self-defense there is grave danger that in the fervor of emotion the main fight will be lost over the question of self-defense.

When my home was bombed in 1955 in Montgomery, many men wanted to retaliate, to place an armed guard on my home. But the issue there was not my life, but whether Negroes would achieve first-class treatment on the city's buses. Had we become distracted by the question of my safety we would have lost the moral offensive and sunk to the level of our oppressors.

I must continue by faith or it is too great a burden to bear and violence, even in self-defense, creates more problems than it solves. Only a refusal to hate or kill can put an end to the chain of violence in the world and lead us toward a community where men can live together without fear. Our goal is to create a beloved community and this will require a qualitative change in our souls as well as a quantitative change in our lives.

STRATEGY FOR CHANGE

The American racial revolution has been a revolution to "get in" rather than to overthrow. We want a share in the American economy, the housing market, the educational system and the social opportunities. This goal itself indicates that a social change in America must be nonviolent.

If one is in search of a better job, it does not help to burn down the factory. If one needs more adequate education, shooting the principal will not help, or if housing is the goal, only building and construction will produce that end. To destroy anything, person or property, can't bring us closer to the goal that we seek.

The nonviolent strategy has been to dramatize the evils of our society in such a way that pressure is brought to bear against those evils by the forces of good will in the community and change is produced.

The student sit-ins of 1960 are a classic illustration of this method. Students were denied the right to eat at a lunch counter, so they deliberately sat down to protest their denial. They were arrested, but this made their parents mad and so they began to close their charge accounts. The students continued to sit in, and this further embarrassed the city, scared away many white shoppers and soon produced an economic threat to the business life of the city. Amid this type of pressure, it is not hard to get people to agree to change.

So far, we have had the Constitution backing most of the demands for change, and this has made our work easier, since we could be sure that the federal courts would usually back up our demonstrations legally. Now we are approaching areas where the voice of the Constitution is not clear. We have left the realm of constitutional rights and we are entering the area of human rights.

The Constitution assured the right to vote, but there is no such assurance of the right to adequate housing, or the right to an adequate income. And yet, in a nation which has a gross national product of 750 billion dollars a year, it is morally right to insist that every person has a decent house, an adequate education and enough money to provide basic necessities for one's family. Achievement of these goals will be a lot more difficult and require much more discipline, understanding, organization and sacrifice.

It so happens that Negroes live in the central city of the major cities of the United States. These cities control the electoral votes of the large states of our nation. This means that though we are only ten percent of the nation's population, we are located in such a key position geographically—the cities of the North and black belts of the South—that we are able to lead a political and moral coalition which can direct the course of the nation. Our position depends upon a lot more than political power, however. It depends upon our ability to marshal moral power as well. As soon as we lose the moral offensive, we are left with only our ten percent of the power of the nation. This is hardly enough to produce any meaningful changes, even within our own communities, for the lines of power control the economy as well and once the flow of money is cut off, progress ceases.

The past three years have demonstrated the power of a committed, morally sound minority to lead the nation. It was the coalition molded through the Birmingham movement which allied the forces of the churches, labor and the academic communities of the nation behind the liberal issues of our time. All of the liberal legislation of the past session of Congress can be credited to this coalition. Even the presence of a vital peace movement and the campus protest against the war in Vietnam can be traced back to the nonviolent action movement led by the Negro. Prior to Birmingham, our campuses were still in a state of shock over the McCarthy era and Congress was caught in the perennial deadlock of southern Democrats and midwestern Republicans. Negroes put the country on the move against the enemies of poverty, slums and inadequate education.

TECHNIQUES OF THE FUTURE

When Negroes marched, so did the nation. The power of the nonviolent march is indeed a mystery. It is always surprising that a few hundred Negroes marching can produce such a reaction across the nation. When marches are carefully organized around well-defined issues, they represent the power which Victor Hugo phrased as the most powerful force in the world, "an idea whose time has come." Marching feet announce that time has come for a given idea. When the idea is a sound one, the cause a just one, and the demonstration a righteous one, change will be forthcoming. But if any of these conditions are not present, the power for change is missing also. A thousand people demonstrating for the right to use heroin would have little effect. By the same token, a group of ten thousand marching in anger against a police station and cussing out the chief of police will do very little to bring respect, dignity and unbiased law enforcement. Such a demonstration would only produce fear and bring about an addition of forces to the station and more oppressive methods by the police.

Marches must continue in the future, and they must be the kind of marches that bring about the desired result. But the march is not a "one shot" victory-producing method. One march is seldom successful, and as my good friend Kenneth Clark points out in *Dark Ghetto,* it can serve merely to let off steam and siphon off the energy which is necessary to produce change. However, when marching is seen as a part of a program to dramatize an evil, to mobilize the forces of good will, and to generate pressure and power for change, marches will continue to be effective.

Our experience is that marches must continue over a period of thirty to forty-five days to produce any meaningful results. They must also be of sufficient size to produce some inconvenience to the forces in power or they go unnoticed. In other words, they must demand the attention of the press, for it is the press which interprets the issue to the community at large and thereby sets in motion the machinery for change.

Along with the march as a weapon for change in our nonviolent arsenal must be listed the boycott. Basic to the philosophy of nonviolence is the refusal to cooperate with evil. There is nothing quite so effective as a refusal to cooperate economically with the forces and institutions which perpetuate evil in our communities.

In the past six months simply by refusing to purchase products from companies which do not hire Negroes in meaningful numbers and in all job categories, the Ministers of Chicago under SCLC's Operation Breadbasket have increased the income of the Negro community by more than two million dollars annually. In Atlanta the Negroes' earning power has been increased by more than twenty million dollars annually over the past three years through a carefully disciplined program of selective buying and negotiations by the Negro minister. This is nonviolence at its peak of power, when it cuts into the profit margin of a business in order to bring about a more just distribution of jobs and opportunities for Negro wage earners and consumers.

But again, the boycott must be sustained over a period of several weeks and months to assure results. This means continuous education of the community in order that support can be maintained. People will work together and sacrifice if they understand clearly why and how this sacrifice will bring about change. We can never assume that anyone understands. It is our job to keep people informed and aware.

Our most powerful nonviolent weapon is, as would be expected, also our most demanding, that is organization. To produce change, people must be organized to work together in units of power. These units might be political, as in the case of voters' leagues and political parties; they may be economic units such as groups of tenants who join forces to form a tenant union or to organize a rent strike; or they may be laboring units of persons who are seeking employment and wage increases.

More and more, the civil rights movement will become engaged in the task of organizing people into permanent groups to protect their own interests and to produce change in their behalf. This is a tedious task which may take years, but the results are more permanent and meaningful.

In the future we will be called upon to organize the unemployed, to unionize the business within the ghetto, to bring tenants together into collective bargaining units and establish cooperatives for purposes of building viable financial institutions within the ghetto that can be controlled by Negroes themselves.

There is no easy way to create a world where men and women can live together, where each has his own job and house and where all children receive as much education as their minds can absorb. But if such a world is created in our lifetime, it will be done in the United States by Negroes and white people of good will. It will be accomplished by persons who have the courage to put an end to suffering by willingly suffering themselves rather than inflict suffering upon others. It will be done by rejecting the racism, materialism and violence that has characterized Western civilization and especially by working toward a world of brotherhood, cooperation and peace.

Ebony 21 (October 1966): 27–30.

16

A Time to Break Silence

(1967)

As the Civil Rights Movement scored significant victories between 1954 and 1965, the United States was gradually escalating its military presence in the civil war raging in Vietnam. A "military-industrial complex," as President Eisenhower had warned, employed its considerable economic and political influence to encourage American military involvements around the globe. Several sages, such as W. E. B. DuBois, had begun pointing out that the victims of American military might were often people of color. From the dropping of the atom bomb on Hiroshima and Nagasaki in 1945 to the Korean War and the war in Vietnam, the United States seemed to find it easier to engage in warfare with people of color than to directly face its nemesis, the Soviet Union. As President Johnson committed more and more troops to fight in Vietnam, people of conscience began to doubt the wisdom of American involvement.

Although Dr. King had spoken against the Vietnam War several times from the pulpit of Ebenezer Baptist Church in Atlanta, Georgia, his first major public declaration came on March 25, 1967, when he led an antiwar demonstration in Chicago with more than 5,000 black and white marchers. But the most controversial and notable statement came just a few days later in New York City. Dr. King delivered this historic address at a meeting of Clergy and Laity Concerned. The meeting was held at the Riverside Church in New York City on April 4, 1967, exactly a year before he was assassinated. This was not the first time he had expressed opposition to the Vietnam War, but it was the first time he linked it to the Civil Rights Movement. And it was the first time that he directly attacked the Johnson administration's war policy.

I COME TO THIS MAGNIFICENT HOUSE OF WORSHIP TONIGHT BECAUSE MY conscience leaves me no other choice. I join with you in this meeting because I am in deepest agreement with the aims and work of the organization which has brought us together: Clergy and Laymen Concerned about Vietnam. The recent statement of your executive committee are the sentiments of my own heart and I found myself in full accord when I read its opening lines: "A time comes when silence is betrayal." That time has come for us in relation to Vietnam.

The truth of these words is beyond doubt but the mission to which they call us is a most difficult one. Even when pressed by the demands of inner truth, men do not easily assume the task of opposing their government's policy, especially in time of war. Nor does the human spirit move without great difficulty against all the apathy of conformist thought within one's own bosom and in the surrounding world. Moreover when the issues at hand seem as perplexed as they often do in the case of this dreadful conflict we are always on the verge of being mesmerized by uncertainty; but we must move on.

Some of us who have already begun to break the silence of the night have found that the calling to speak is often a vocation of agony, but we must speak. We must speak with all the humility that is appropriate to our limited vision, but we must speak. And we must rejoice as well, for surely this is the first time in our nation's history that a significant number of its religious leaders have chosen to move beyond the prophesying of smooth patriotism to the high grounds of a firm dissent based upon the mandates of conscience and the reading of history. Perhaps a new spirit is rising among us. If it is, let us trace its movement well and pray that our own inner being may be sensitive to its guidance, for we are deeply in need of a new way beyond the darkness that seems so close around us.

Over the past two years, as I have moved to break the betrayal of my own silences and to speak from the burnings of my own heart, as I have called for radical departures from the destruction of Vietnam, many persons have questioned me about the wisdom of my path. At the heart of their concerns this query has often loomed large and loud: Why are *you* speaking about war, Dr. King? Why are *you* joining the voices of dissent? Peace and civil rights don't mix, they say. Aren't you hurting the cause of your people, they ask? And when I hear them, though I often understand the source of their concern, I am nevertheless greatly saddened, for such questions mean that the inquirers have not really known me, my commitment or my calling. Indeed, their questions suggest that they do not know the world in which they live.

In the light of such tragic misunderstandings, I deem it of signal importance to try to state clearly, and I trust concisely, why I believe that the path from Dexter Avenue Baptist Church—the church in Montgomery, Alabama, where I began my pastorate—leads clearly to this sanctuary tonight.

I come to this platform tonight to make a passionate plea to my beloved nation. This speech is not addressed to Hanoi or to the National Liberation Front. It is not addressed to China or to Russia.

Nor is it an attempt to overlook the ambiguity of the total situation and the need for a collective solution to the tragedy of Vietnam. Neither is it an attempt to make North Vietnam or the National Liberation Front paragons of virtue, nor to overlook the role they can play in a successful resolution of the problem. While they both may have justifiable reason to be suspicious of the good faith of the United States, life and history give eloquent testimony to the fact that conflicts are never resolved without trustful give and take on both sides.

Tonight, however, I wish not to speak with Hanoi and the NLF, but rather to my fellow Americans who, with me, bear the greatest responsibility in ending a conflict that has exacted a heavy price on both continents.

IMPORTANCE OF VIETNAM

Since I am a preacher by trade, I suppose it is not surprising that I have seven major reasons for bringing Vietnam into the field of my moral vision. There is at the outset a very obvious and almost facile connection between the war in Vietnam and the struggle I, and others, have been waging in America. A few years ago there was a shining moment in that struggle. It seemed as if there was a real promise of hope for the poor—both black and white—through the poverty program. There were experiments, hopes, new beginnings. Then came the buildup in Vietnam and I watched the program broken and eviscerated as if it were some idle political plaything of a society gone mad on war, and I knew that America would never invest the necessary funds or energies in rehabilitation of its poor so long as adventures like Vietnam continued to draw men and skills and money like some demonic destructive suction tube. So I was increasingly compelled to see the war as an enemy of the poor and to attack it as such.

Perhaps the more tragic recognition of reality took place when it became clear to me that the war was doing far more than devastating the hopes of the poor at home. It was sending their sons and their brothers and their husbands to fight and to die in extraordinarily high proportions relative to the rest of the population. We were taking the black young men who had been crippled by our society and sending them eight thousand miles away to guarantee liberties in Southeast Asia which they had not found in southwest Georgia and East Harlem. So we have been repeatedly faced with the cruel irony of watching Negro and white boys on TV screens as they kill and die together for a nation that has been unable to seat them together in the same schools. So we watch them in brutal solidarity burning the huts of a poor village, but we realize that they would never live on the same block in Detroit. I could not be silent in the face of such cruel manipulation of the poor.

My third reason moves to an even deeper level of awareness, for it grows out of my experience in the ghettos of the North over the last three years—especially the last three summers. As I have walked among the desperate, rejected and angry young men I have told them that Molotov cocktails and rifles would not solve their problems. I have tried to offer them my deepest compassion while maintaining my conviction that social change comes most meaningfully through nonviolent action. But they asked—and rightly so—what about Vietnam? They asked if our own nation wasn't using massive doses of violence to solve its problems, to bring about the changes it wanted. Their questions hit home, and I knew that I could never again raise my voice against the violence of the oppressed in the ghettos without having first spoken clearly to the greatest purveyor of violence in the world today—my own government. For the sake of those boys, for the sake of this government, for the sake of the hundreds of thousands trembling under our violence, I cannot be silent.

For those who ask the question, "Aren't you a civil rights leader?" and thereby mean to exclude me from the movement for peace, I have this further answer. In 1957 when a group of us formed the Southern Christian Leadership Conference, we chose as our motto: "To save the soul of America." We were convinced that we could not limit our vision to certain rights for black people, but instead affirmed the conviction that America would never be free or saved from itself unless the descendants of its slaves were loosed completely from the shackles they still wear. In a way we were agreeing with Langston Hughes, that black bard of Harlem, who had written earlier:

> O, yes,
> I say it plain,
> America never was America to me,
> And yet I swear this oath—
> America will be!

Now, it should be incandescently clear that no one who has any concern for the integrity and life of America today can ignore the present war. If America's soul becomes totally poisoned, part of the autopsy must read Vietnam. It can never be saved so long as it destroys the deepest hopes of men the world over. So it is that those of us who are yet determined that America *will* be are led down the path of protest and dissent, working for the health of our land.

As if the weight of such a commitment to the life and health of America were not enough, another burden of responsibility was placed upon me in 1964; and I cannot forget that the Nobel Prize for Peace was also a commission—a commission to work harder than I had ever worked before "the brotherhood of man." This is a calling that takes me beyond national allegiances, but even if it were not present I would yet have to live with the meaning of my commitment to the ministry of Jesus Christ. To me the relationship of this ministry to the making of peace is so obvious that I sometimes marvel at those who ask me why I am speaking against the war. Could it be that they do not know that the good news was meant for all men—for Communist and capitalist, for their children and ours, for black and for white, for revolutionary and conservative? Have they forgotten that my ministry is in obedience to the one who loved his enemies so fully that he died for them? What then can I say to the "Vietcong" or to Castro or to Mao as a faithful minister of this one? Can I threaten them with death or must I not share with them my life?

Finally, as I try to delineate for you and for myself the road that leads from Montgomery to this place I would have offered all that was most valid if I simply said that I must be true to my conviction that I share with all men the calling to be a son of the living God. Beyond the calling of race or nation or creed is this vocation of sonship and brotherhood, and because I believe that the Father is deeply concerned especially for his suffering and helpless and outcast children, I come tonight to speak for them.

This I believe to be the privilege and the burden of all of us who deem ourselves bound by allegiances and loyalties which are broader and deeper than nationalism and which go beyond our nation's self-defined goals and positions. We are called to speak for the weak, for the voiceless, for victims of our nation and for those it calls enemy, for no document from human hands can make these humans any less our brothers.

STRANGE LIBERATORS

And as I ponder the madness of Vietnam and search within myself for ways to understand and respond to compassion my mind goes constantly to the people of that peninsula. I speak now not of the soldiers of each side, not of the junta in Saigon, but simply of the people who have been living under the curse of war for almost three continuous decades now. I think of them too because it is clear to me that there will be no meaningful solution there until some attempt is made to know them and hear their broken cries.

They must see Americans as strange liberators. The Vietnamese people proclaimed their own independence in 1945 after a combined French and Japanese occupation, and before the Communist revolution in China. They were led by Ho Chi Minh. Even though they quoted the American Declaration of Independence in their own document of freedom, we refused to recognize them. Instead, we decided to support France in its reconquest of her former colony.

Our government felt then that the Vietnamese people were not "ready" for independence, and we again fell victim to the deadly Western arrogance that has poisoned the international atmosphere for so long. With that tragic decision we rejected a revolutionary government seeking self-determination, and a government that had been established not by China (for whom the Vietnamese have no great love) but by clearly indigenous forces that included some Communists. For the peasants this new government meant real land reform, one of the most important needs in their lives.

For nine years following 1945 we denied the people of Vietnam the right of independence. For nine years we vigorously supported the French in their abortive effort to recolonize Vietnam.

Before the end of the war we were meeting eighty percent of the French war costs. Even before the French were defeated at Dien Bien Phu, they began to despair of the reckless action, but we did not. We encouraged them with our huge financial and military supplies to continue the war even after they had lost the will. Soon we would be paying almost the full costs of this tragic attempt at recolonization.

After the French were defeated it looked as if independence and land reform would come again through the Geneva agreements. But instead there came the United States, determined that Ho should not unify the temporarily divided nation, and the peasants watched again as we supported one of the most vicious modern dictators—our chosen man, Premier Diem. The peasants watched and cringed as Diem ruthlessly routed out all opposition, supported their extortionist landlords and refused even to discuss reunification with the north. The peasants watched as all this was presided over by U.S. influence and then by increasing numbers of U.S. troops who came to help quell the insurgency that Diem's methods had aroused. When Diem was overthrown they may have been happy, but the long line of military dictatorships seemed to offer no real change—especially in terms of their need for land and peace.

The only change came from America as we increased our troop commitments in support of governments which were singularly corrupt, inept and without popular support. All the while the people read our leaflets and received regular promises of peace and democracy—and land reform. Now they languish under our bombs and consider us—not their fellow Vietnamese—the real enemy. They move sadly and apathetically as we herd them off the land of their fathers into concentration camps where minimal social needs are rarely met. They know they must move or be destroyed by our bombs. So they go—primarily women and children and the aged.

They watch as we poison their water, as we kill a million acres of their crops. They must weep as the bulldozers roar through their areas preparing to destroy the precious trees. They wander into the hospitals, with at least twenty casualties from American firepower for one "Vietcong"-inflicted injury. So far we may have killed a million of them—mostly children. They wander into the towns and see thousands of the children, homeless, without clothes, running in packs on the streets like animals. They see the children degraded by our soldiers as they beg for food. They see the children selling their sisters to our soldiers, soliciting for their mothers.

What do the peasants think as we ally ourselves with the landlords and as we refuse to put any action into our many words concerning land reform? What do they think as we test out our latest weapons on them, just as the Germans tested out new medicine and new tortures in the concentration camps of Europe? Where are the roots of the independent Vietnam we claim to be building? Is it among these voiceless ones?

We have destroyed their two most cherished institutions: the family and the village. We have destroyed their land and their crops. We have cooperated in the crushing of the nation's only non-Communist revolutionary political force—the unified Buddhist church. We have supported the enemies of the peasants of Saigon. We have corrupted their women and children and killed their men. What liberators!

Now there is little left to build on—save bitterness. Soon the only solid physical foundations remaining will be found at our military bases and in the concrete of the concentration camps we call fortified hamlets. The peasants may well wonder if we plan to build our new Vietnam on such grounds as these? Could we blame them for such thoughts? We must speak for them and raise the questions they cannot raise. These too are our brothers.

Perhaps the more difficult but no less necessary task is to speak for those who have been designated as our enemies. What of the National Liberation Front—that strangely anonymous group we call VC or Communists? What must they think of us in America when they realize that we permitted the repression and cruelty of Diem which helped to bring them into being as a resistance group in the south? What do they think of our condoning the violence which led to their own taking up of arms? How can they believe in our integrity when now we speak of "aggression from the north" as if there were nothing more essential to the war? How can they trust us when now we charge them with violence after the murderous reign of Diem and charge them with violence while we pour every new weapon of death into their land? Surely we must understand their feelings even if we do not condone their actions. Surely we must see that the men we supported pressed them to their violence. Surely we must see that our own computerized plans of destruction simply dwarf their greatest acts.

How do they judge us when our officials know that their membership is less than twenty-five percent Communist and yet insist on giving them the blanket name? What must they be thinking when they know that we are aware of their control of major sections of Vietnam and yet we appear ready to allow national elections in which this highly organized political parallel government will have no part? They ask how we can speak of free elections when the Saigon press is censored and controlled by the military junta. And they are surely right to wonder what kind of new government we plan to help form without them—the only party in real touch with the peasants. They question our political goals and they deny the reality of a peace settlement from which they will be excluded. Their questions are frighteningly relevant. Is our nation planning to build on political myth again and then shore it up with the power of new violence?

Here is the true meaning and value of compassion and nonviolence when it helps us to see the enemy's point of view, to hear his questions, to know his assessment of ourselves. For from his view we may indeed see the basic weaknesses of our own condition, and if we are mature, we may learn and grow and profit from the wisdom of the brothers who are called the opposition.

So, too, with Hanoi. In the north, where our bombs now pummel the land, and our mines endanger the waterways, we are met by a deep but understandable mistrust. To speak for them is to explain this lack of confidence in Western words, and especially their distrust of American intentions now. In Hanoi are the men who led the nation to independence against the Japanese and the French, the men who sought membership in the French commonwealth and were betrayed by the weakness of Paris and the willfulness of the colonial armies. It was they who led a second struggle against French domination at tremendous costs, and then were persuaded to give up the land they controlled between the thirteenth and seventeenth parallel as a temporary measure at Geneva. After 1954 they watched us conspire with Diem to prevent elections which would have surely brought Ho Chi Minh to power over a united Vietnam, and they realized they had been betrayed again.

When we ask why they do not leap to negotiate, these things must be remembered. Also it must be clear that the leaders of Hanoi considered the presence of American troops in support of the Diem regime to have been the initial military breach of the Geneva agreements concerning foreign troops, and they remind us that they did not begin to send in any large number of supplies or men until American forces had moved into the tens of thousands.

Hanoi remembers how our leaders refused to tell us the truth about the earlier North Vietnamese overtures for peace, how the president claimed that none existed when they had clearly been made. Ho Chi Minh has watched as America has spoken of peace and built up its forces, and now he has surely heard of the increasing international rumors of American plans for an invasion of the north. He knows the bombing and shelling and mining we are doing are part of traditional pre-invasion strategy. Perhaps only his sense of humor and of irony can save him when he hears the most powerful nation of the world speaking of aggression as it drops thousands of bombs on a poor weak nation more than eight thousand miles away from its shores.

At this point I should make it clear that while I have tried in these last few minutes to give a voice to the voiceless on Vietnam and to understand the arguments of those who are called enemy, I am as deeply concerned about our troops there as anything else. For it occurs to me that what we are submitting them to in Vietnam is not simply the brutalizing process that goes on in any war where armies face each other and seek to destroy. We are adding cynicism to the process of death, for they must know after a short period there that none of the things we claim to be fighting for are really involved. Before long they must know that their government has sent them into a struggle among Vietnamese, and the more sophisticated surely realize that we are on the side of the wealthy and the secure while we create a hell for the poor.

Somehow this madness must cease. We must stop now. I speak as a child of God and brother to the suffering poor of Vietnam. I speak for those whose land is being laid waste, whose homes are being destroyed, whose culture is being subverted. I speak for the poor of America who are paying the double price of smashed hopes at home and death and corruption in Vietnam. I speak as a citizen of the world, for the world as it stands aghast at the path we have taken. I speak as an American to the leaders of my own nation. The great initiative in this war is ours. The initiative to stop it must be ours.

This is the message of the great Buddhist leaders of Vietnam. Recently one of them wrote these words: *Each day the war goes on the hatred increases in the heart of the Vietnamese and in the hearts of those of humanitarian instinct. The Americans are forcing even their friends into becoming their enemies. It is curious that the Americans, who calculate so carefully on the possibilities of military victory, do not realize that in the process they are incurring deep psychological and political defeat. The image of America will never again be the image of revolution, freedom and democracy, but the image of violence and militarism.*

If we continue there will be no doubt in my mind and in the mind of the world that we have no honorable intentions in Vietnam. It will become clear that our minimal expectation is to occupy it as an American colony and men will not refrain from thinking that our maximum hope is to goad China into a war so that we may bomb her nuclear installations. If we do not stop our war against the people of Vietnam immediately the world will be left with no other alternative than to see this as some horribly clumsy and deadly game we have decided to play.

The world now demands a maturity of America that we may not be able to achieve. It demands that we admit that we have been wrong from the beginning of our adventure in Vietnam, that we have been detrimental to the life of the Vietnamese people. The situation is one in which we must be ready to turn sharply from our present ways.

In order to atone for our sins and errors in Vietnam, we should take the initiative in bringing a halt to this tragic war. I would like to suggest five concrete things that our government should do immediately to begin the long and difficult process of extricating ourselves from this nightmarish conflict:

1. *End all bombing in North and South Vietnam.*
2. *Declare a unilateral cease-fire in the hope that such action will create the atmosphere for negotiation.*
3. *Take immediate steps to prevent other battlegrounds in Southeast Asia by curtailing our military buildup in Thailand and our interference in Laos.*
4. *Realistically accept the fact that the National Liberation Front has substantial support in South Vietnam and must thereby play a role in any meaningful negotiations and in any future Vietnam government.*
5. *Set a date that we will remove all foreign troops from Vietnam in accordance with the 1954 Geneva agreement.*

Part of our ongoing commitment might well express itself in an offer to grant asylum to any Vietnamese who fears for his life under a new regime which included the Liberation Front. Then we must make what reparations we can for the damage we have done. We must provide the medical aid that is badly needed, making it available in this country if necessary.

PROTESTING THE WAR

Meanwhile we in the churches and synagogues have a continuing task while we urge our government to disengage itself from a disgraceful commitment. We must continue to raise our voices if our nation persists in its perverse ways in Vietnam. We must be prepared to match actions with words by seeking out every creative means of protest possible.

As we counsel young men concerning military service we must clarify for them our nation's role in Vietnam and challenge them with the alternative of conscientious objection. I am pleased to say that this is the path now being chosen by more than seventy students at my own alma mater, Morehouse College, and I recommend it to all who find the American course in Vietnam a dishonorable and unjust one. Moreover I would encourage all ministers of draft age to give up their ministerial exemptions and seek status as conscientious objectors. These are the times for real choices and not false ones. We are at the moment when our lives must be placed on the line if our nation is to survive its own folly. Every man of humane convictions must decide on the protest that best suits his convictions, but we must all protest.

There is something seductively tempting about stopping there and sending us all off on what in some circles has become a popular crusade against the war in Vietnam. I say we must enter the struggle, but I wish to go on now to say something even more disturbing. The war in Vietnam is but a symptom of a far deeper malady within the American spirit, and if we ignore this sobering reality we will find ourselves organizing clergy- and laymen-concerned committees for the next generation. They will be concerned about Guatemala and Peru. They will be concerned about Thailand and Cambodia. They will be concerned about Mozambique and South Africa. We will be marching for these and a dozen other names and attending rallies without end unless there is a significant and profound change in American life and policy. Such thoughts take us beyond Vietnam, but not beyond our calling as sons of the living God.

In 1957 a sensitive American official overseas said that it seemed to him that our nation was on the wrong side of a world revolution. During the past ten years we have seen emerge a pattern of suppression which now has justified the presence of U.S. military "advisors" in Venezuela. This need to maintain social stability for our investments accounts for the counter-revolutionary action of American forces in Guatemala. It tells why American helicopters are being used against guerrillas in Colombia and why American napalm and green beret forces have already been active against rebels in Peru. It is with such activity in mind that the words of the late John F. Kennedy come back to haunt us. Five years ago he said, "Those who make peaceful revolution impossible will make violent revolution inevitable."

Increasingly, by choice or by accident, this is the role our nation has taken—the role of those who make peaceful revolution impossible by refusing to give up the privileges and the pleasures that come from the immense profits of overseas investment.

I am convinced that if we are to get on the right side of the world revolution, we as a nation must undergo a radical revolution of values. We must rapidly begin the shift from a "thing-oriented" society to a "person-oriented" society. When machines and computers, profit motives and property rights are considered more important than people, the giant triplets of racism, materialism, and militarism are incapable of being conquered.

A true revolution of values will soon cause us to question the fairness and justice of many of our past and present policies. On the one hand we are called to play the good Samaritan on life's roadside; but that will be only an initial act. One day we must come to see that the whole Jericho road must be transformed so that men and women will not be constantly beaten and robbed as they make their journey on life's highway. True compassion is more than flinging a coin to a beggar; it is not haphazard and superficial. It comes to see that an edifice which produces beggars needs restructuring. A true revolution of values will soon look uneasily on the glaring contrast of poverty and wealth. With righteous indignation, it will look across the seas and see individual capitalists of the West investing huge sums of money in Asia, Africa and South America, only to take the profits out with no concern for the social betterment of the countries, and say: "This is not just." It will look at our alliance with the landed gentry of Latin America and say: "This is not just." The Western arrogance of feeling that it has everything to teach others and nothing to learn from them is not just. A true revolution of values will lay hands on the world order and say of war: "This way of settling differences is not just." This business of burning human beings with napalm, of filling our nation's homes with orphans and widows, of injecting poisonous drugs of hate into veins of peoples normally humane, of sending men home from dark and bloody battlefields physically handicapped and psychologically deranged, cannot be reconciled with wisdom, justice and love. A nation that continues year after year to spend more money on military defense than on programs of social uplift is approaching spiritual death.

America, the richest and most powerful nation in the world, can well lead the way in this revolution of values. There is nothing, except a tragic death wish, to prevent us from reordering our priorities, so that the pursuit of peace will take precedence over the pursuit of war. There is nothing to keep us from molding a recalcitrant status quo with bruised hands until we have fashioned it into a brotherhood.

This kind of positive revolution of values is our best defense against communism. War is not the answer. Communism will never be defeated by the use of atomic bombs or nuclear weapons. Let us not join those who shout war and through their misguided passions urge the United States to relinquish its participation in the United Nations. These are days which demand wise restraint and calm reasonableness. We must not call everyone a Communist or an appeaser who advocates the seating of Red China in the United Nations and who recognizes that hate and hysteria are not the final answers to the problem of these turbulent days. We must not engage in a negative anti-communism, but rather in a positive thrust for democracy, realizing that our greatest defense against communism is to take offensive action in behalf of justice. We must with positive action seek to remove those conditions of poverty, insecurity and injustice which are the fertile soil in which the seed of communism grows and develops.

THE PEOPLE ARE IMPORTANT

These are revolutionary times. All over the globe men are revolting against old systems of exploitation and oppression and out of the wombs of a frail world new systems of justice and equality are being born. The shirtless and barefoot people of the land are rising up as never before. "The people who sat in darkness have seen a great light." We in the West must support these revolutions. It is a sad fact that, because of comfort, complacency, a morbid fear of communism, and our proneness to adjust to injustice, the Western nations that initiated so much of the revolutionary spirit of the modern world have now become the arch anti-revolutionaries. This has driven many to feel that only Marxism has the revolutionary spirit. Therefore, communism is a judgment against our failure to make democracy real and follow through on the revolutions that we initiated. Our only hope today lies in our ability to recapture the revolutionary spirit and go out into a sometimes hostile world declaring eternal hostility to poverty, racism, and militarism. With this powerful commitment we shall boldly challenge the status quo and unjust mores and thereby speed the day when "every valley shall be exalted, and every mountain and hill shall be made low, and the crooked shall be made straight and the rough places plain."

A genuine revolution of values means in the final analysis that our loyalties must become ecumenical rather than sectional. Every nation must now develop an overriding loyalty to mankind as a whole in order to preserve the best in their individual societies.

This call for a world-wide fellowship that lifts neighborly concern beyond one's tribe, race, class and nation is in reality a call for an all-embracing and unconditional love for all men. This oft misunderstood and misinterpreted concept—so readily dismissed by the Nietzsches of the world as a weak and cowardly force—has now become an absolute necessity for the survival of man. When I speak of love I am not speaking of some sentimental and weak response. I am speaking of that force which all of the great religions have seen as the supreme unifying principle of life. Love is somehow the key that unlocks the door which leads to ultimate reality. This Hindu-Moslem-Christian-Jewish-Buddhist belief about ultimate reality is beautifully summed up in the first epistle of Saint John:

> Let us love one another; for love is God and everyone that loveth is born of God and knoweth God. He that loveth not knoweth not God; for God is love. If we love one another God dwelleth in us, and his love is perfected in us.

Let us hope that this spirit will become the order of the day. We can no longer afford to worship the god of hate or bow before the altar of retaliation. The oceans of history are made turbulent by the ever-rising tides of hate. History is cluttered with the wreckage of nations and individuals that pursued this self-defeating path of hate. As Arnold Toynbee says: "Love is the ultimate force that makes for the saving choice of life and good against the damning choice of death and evil. Therefore the first hope in our inventory must be the hope that love is going to have the last word."

We are now faced with the fact that tomorrow is today. We are confronted with the fierce urgency of now. In this unfolding conundrum of life and history there is such a thing as being too late. Procrastination is still the thief of time. Life often leaves us standing bare, naked and dejected with a lost opportunity. The "tide in the affairs of men" does not remain at the flood; it ebbs. We may cry out desperately for time to pause in her passage, but time is deaf to every plea and rushes on. Over the bleached bones and jumbled residue of numerous civilizations are written the pathetic words: "Too late." There is an invisible book of life that faithfully records our vigilance or our neglect. "The moving finger writes, and having writ moves on...." We still have a choice today; nonviolent coexistence or violent co-annihilation.

We must move past indecision to action. We must find new ways to speak for peace in Vietnam and justice throughout the developing world—a world that borders on our doors. If we do not act we shall surely be dragged down the long dark and shameful corridors of time reserved for those who possess power without compassion, might without morality, and strength without sight.

Now let us begin. Now let us rededicate ourselves to the long and bitter—but beautiful—struggle for a new world. This is the calling of the sons of God, and our brothers wait eagerly for our response. Shall we say the odds are too great? Shall we tell them the struggle is too hard? Will our message be that the forces of American life militate against their arrival as full men, and we send our deepest regrets? Or will there be another message, of longing, of hope, of solidarity with their yearnings, of commitment to their cause, whatever the cost? The choice is ours, and though we might prefer it otherwise we *must* choose in this crucial moment of human history.

As that noble bard of yesterday, James Russell Lowell, eloquently stated:

> Once to every man and nation
> Comes the moment to decide,
> In the strife of truth and falsehood,
> For the good or evil side;
> Some great cause, God's new Messiah,
> Off'ring each the bloom or blight,

And the choice goes by forever
Twixt that darkness and that light.

Though the cause of evil prosper,
Yet 'tis truth alone is strong;
Though her portion be the scaffold,
And upon the throne be wrong:
Yet that scaffold sways the future,
And behind the dim unknown,
Standeth God within the shadow
Keeping watch above his own.

Freedomways 7 (Spring 1967): 103–17.

17
Black Power Defined

(1967)

The following article, which appeared in the *New York Times Magazine,* is a synopsis of Dr. King's understanding of the new wave of African American nationalism then being embraced by many within the black community. His book *Where Do We Go from Here: Chaos or Community?* (New York: Harper & Row, 1967) contains a more expanded discussion of this movement. Dr. King embraced the basic political and social agenda of black power advocates, but strongly condemned their endorsement of revolutionary violence and black separatism.

The Black Power Movement advocated the need for people of African American descent to become more conscious of African history and culture. Its adherents argued that African people in the Western Hemisphere were united by the experience of enslavement and colonialism, and that they should come to see themselves as victims of western industrial capitalism rather than its beneficiaries. The call for African American solidarity based on a common history and culture became the broad strategy of the Black Power Movement. Many of the ideas and programs of this movement had been advanced by earlier African American leaders such as Frederick Douglass (1817–1895), Alexander Crummell (1819–1898), Booker T. Washington (1856–1915), Ida Bell Wells-Barnett (1862–1931), William E. B. Du Bois (1868–1963), and Marcus Garvey (1887–1940). All of these leaders in one way or another argued the need for African American self-respect and self-help. But the advocacy for these positions and endeavors makes little sense outside of the context of the history of white racism and oppression. This history included, as stated in the introduction to this book, a sustained attempt to use racist caricatures of African people and various forms of social, economic and psychological terrorism to deny African Americans basic rights guaranteed to every citizen by the Constitution of the United States.

WHEN A PEOPLE ARE MIRED IN OPPRESSION, THEY REALIZE DELIVERANCE only when they have accumulated the power to enforce change. The powerful never lose opportunities—they remain available to them. The powerless, on the other hand, never experience opportunity—it is always arriving at a later time.

The nettlesome task of Negroes today is to discover how to organize our strength into compelling power so that government cannot elude our demands. We must develop, from strength, a situation in which the government finds it wise and prudent to collaborate with us. It would be the height of naiveté to wait passively until the administration had somehow been infused with such blessings of good will that it implored us for our programs.

We must frankly acknowledge that in past years our creativity and imagination were not employed in learning how to develop power. We found a method in nonviolent protest that worked, and we employed it enthusiastically. We did not have leisure to probe for a deeper understanding of its laws and lines of development. Although our actions were bold and crowned with successes, they were substantially improvised and spontaneous. They attained the goals set for them but carried the blemishes of our inexperience.

This is where the civil rights movement stands today. Now we must take the next major step of examining the levers of power which Negroes must grasp to influence the course of events.

In our society power sources can always finally be traced to ideological, economic and political forces.

In the area of *ideology,* despite the impact of the works of a few Negro writers on a limited number of white intellectuals, all too few

Negro thinkers have exerted an influence on the main currents of American thought. Nevertheless, Negroes have illuminated imperfections in the democratic structure that were formerly only dimly perceived, and have forced a concerned reexamination of the true meaning of American democracy. As a consequence of the vigorous Negro protest, the whole nation has for a decade probed more searchingly the essential nature of democracy, both economic and political. By taking to the streets and there giving practical lessons in democracy and its defaults, Negroes have decisively influenced white thought.

Lacking sufficient access to television, publications and broad forums, Negroes have had to write their most persuasive essays with the blunt pen of marching ranks. The many white political leaders and well-meaning friends who ask Negro leadership to leave the streets may not realize that they are asking us effectively to silence ourselves. More white people learned more about the shame of America, and finally faced some aspects of it, during the years of nonviolent protest than during the century before. Nonviolent direct action will continue to be a significant source of power until it is made irrelevant by the presence of justice.

The *economic* highway to power has few entry lanes for Negroes. Nothing so vividly reveals the crushing impact of discrimination and the heritage of exclusion as the limited dimensions of Negro business in the most powerful economy in the world. America's industrial production is half of the world's total, and within it the production of Negro business is so small that it can scarcely be measured on any definable scale.

Yet in relation to the Negro community the value of Negro business should not be underestimated. In the internal life of the Negro society it provides a degree of stability. Despite formidable obstacles it has developed a corps of men of competence and organizational discipline who constitute a talented leadership reserve, who furnish inspiration and who are a resource for the development of programs and planning. They are a strength among the weak though they are weak among the mighty.

There exist two other areas, however, where Negroes can exert substantial influence on the broader economy. As employees and consumers, Negro numbers and their strategic disposition endow them with a certain bargaining strength.

Within the ranks of organized labor there are nearly two million Negroes, and they are concentrated in key industries. In the truck transportation, steel, auto and food industries, which are the backbone of the nation's economic life, Negroes make up nearly twenty percent of the organized work force, although they are only ten percent of the general population. This potential strength is magnified further by the fact of their unity with millions of white workers in these occupations. As co-workers there is a basic community of interest that transcends many of the ugly divisive elements of traditional prejudice. There are undeniably points of friction, for example, in certain housing and education questions. But the severity of the abrasions is minimized by the more commanding need for cohesion in union organizations.

The union record in relation to Negro workers is exceedingly uneven, but potential for influencing union decisions still exists. In many of the larger unions the white leadership contains some men of ideals and many more who are pragmatists. Both groups find they are benefited by a constructive relationship to their Negro membership. For those compelling reasons, Negroes, who are almost wholly a working people, cannot be casual toward the union movement. This is true even though some unions remain uncontestably hostile.

In days to come, organized labor will increase its importance in the destinies of Negroes. Negroes pressed into the proliferating service occupations—traditionally unorganized and with low wages and long hours—need union protection, and the union movement needs their membership to maintain its relative strength in the whole society. On this new frontier Negroes may well become the pioneers that they were in the early organizing days of the thirties.

To play our role fully as Negroes we will also have to strive for enhanced representation and influence in the labor movement. Our young people need to think of union careers as earnestly as they do of business careers and professions. They could do worse than emulate A. Philip Randolph, who rose to the executive council of the AFL-CIO and became a symbol of the courage, compassion and integrity of an enlightened labor leader.

Indeed, the question may be asked why we have produced only one Randolph in nearly half a century. Discrimination is not the whole answer. We allowed ourselves to accept middle-class prejudices against the labor movement. Yet this is one of those fields in which higher education is not a requirement for high office. In shunning it, we have lost an opportunity. Let us try to regain it now, at a time when the joint forces of Negroes and labor may be facing a historic task of social reform.

The other economic lever available to the Negro is as a consumer. The Southern Christian Leadership Council has pioneered in developing mass boycott movements in a frontal attack on discrimination. In Birmingham it was not the marching alone that brought about integration of public facilities in 1963. The downtown business establishments suffered for weeks under our almost unbelievably effective boycott. The significant percentage of their sales that vanished, the ninety-eight percent of their Negro customers who stayed home, educated them forcefully to the dignity of the Negro as a consumer.

Later we crystallized our experiences in Birmingham and elsewhere and developed a department in SCLC called Operation Breadbasket. This has as its primary aim the securing of more and better jobs for the Negro people. It calls on the Negro community to support those businesses that will give a fair share of jobs to Negroes and to withdraw its support from those businesses that have discriminatory policies.

Operation Breadbasket is carried out mainly by clergymen. First, a team of ministers calls on the management of a business in the community to request basic facts on the company's total number of employees, the number of Negro employees, the departments or job classifications in which all employees are located, and the salary ranges for each category. The team then returns to the steering committee to evaluate the data and to make a recommendation concerning the number of new and upgraded jobs that should be requested. Then the team transmits the request to the management to hire or upgrade a specified number of "qualifiable" Negroes within a reasonable period of time. If negotiations on this request break down, the step of real power and pressure is taken: a massive call for economic withdrawal from the company's product and accompanying demonstrations if necessary.

At present SCLC has Operation Breadbasket functioning in some twelve cities, and the results have been remarkable. In Atlanta, for instance, the Negroes' earning power has been increased by more than twenty million dollars annually over the past three years through a carefully disciplined program of selective buying and negotiation by the Negro ministers. During the last eight months in Chicago, Operation Breadbasket successfully completed negotiations with three major industries: milk, soft drinks and chain grocery stores. Four of the companies involved concluded reasonable agreements only after short "don't buy" campaigns. Seven other companies were able to make the requested changes across the conference table, without necessitating a boycott. Two other companies, after providing their employment information to the ministers, were sent letters of commendation for their healthy equal-employment practices. The net results add up to approximately eight hundred new and upgraded jobs for Negro employees, worth a little over seven million dollars in new annual income for Negro families. In Chicago we have recently added a new dimension to Operation Breadbasket. Along with requesting new job opportunities, we are now requesting that businesses with stores in the ghetto deposit the income for those establishments in Negro-owned banks, and that Negro-owned products be placed on the counters of all their stores. In this way we seek to stop the drain of resources out of the ghetto with nothing remaining there for its rehabilitation.

The final major area of untapped power for the Negro is in the *political* arena. Higher Negro birth rates and increasing Negro migration, along with the exodus of the white population to the suburbs, are producing fast-gathering Negro majorities in the large cities. This changing composition of the cities has political significance. Particularly in the North, the large cities substantially determine the political destiny of the state. These states, in turn, hold the dominating electoral votes in presidential contests. The future of the Democratic party, which rests so heavily on its coalition of urban minorities, cannot be assessed without taking into account which way the Negro vote turns. The wistful hopes of the Republican party for large-city influence will also be decided not in the boardrooms of great corporations but in the teeming ghettos.

The growing Negro vote in the South is another source of power. As it weakens and enfeebles the dixiecrats, by concentrating its blows against them, it undermines the congressional coalition of southern reactionaries and their northern Republican colleagues.

That coalition, which has always exercised a disproportionate power in Congress by controlling its major committees, will lose its ability to frustrate measures of social advancement and to impose its perverted definition of democracy on the political thought of the nation.

The Negro vote at present is only a partially realized strength. It can still be doubled in the South. In the North even where Negroes are registered in equal proportion to whites, they do not vote in the same proportions. Assailed by a sense of futility, Negroes resist participating in empty ritual. However, when the Negro citizen learns that united and organized pressure can achieve measurable results, he will make his influence felt. Out of this conscious act, the political power of the aroused minority will be enhanced and consolidated.

We have many assets to facilitate organization. Negroes are almost instinctively cohesive. We band together readily, and against white hostility we have an intense and wholesome loyalty to each other. We are acutely conscious of the need, and sharply sensitive to the importance, of defending our own. Solidarity is a reality in Negro life, as it always has been among the oppressed.

On the other hand, Negroes are capable of becoming competitive, carping and, in an expression of self-hate, suspicious and intolerant of each other. A glaring weakness in Negro life is lack of sufficient mutual confidence and trust.

Negro leaders suffer from this interplay of solidarity and divisiveness, being either exalted excessively or grossly abused. Some of these leaders suffer from an aloofness and absence of faith in their people. The white establishment is skilled in flattering and cultivating emerging leaders. It presses its own image on them and finally, from imitation of manners, dress and style of living, a deeper strain of corruption develops. This kind of Negro leader acquires the white man's contempt for the ordinary Negro. He is often more at home with the middle-class white than he is among his own people. His language changes, his location changes, his income changes, and ultimately he changes from the representative of the Negro to the white man into the white man's representative to the Negro. The tragedy is that too often he does not recognize what has happened to him.

I learned a lesson many years ago from a report of two men who flew to Atlanta to confer with a Negro civil rights leader at the airport. Before they could begin to talk, the porter sweeping the floor drew the local leader aside to talk about a matter that troubled him. After fifteen minutes had passed, one of the visitors said bitterly to his companion, "I am just too busy for this kind of nonsense. I haven't come a thousand miles to sit and wait while he talks to a porter."

The other replied, "When the day comes that he stops having time to talk to a porter, on that day I will not have the time to come one mile to see him."

We need organizations that are permeated with mutual trust, incorruptibility and militancy. Without this spirit we may have numbers but they will add up to zero. We need organizations that are responsible, efficient and alert. We lack experience because ours is a history of disorganization. But we will prevail because our need for progress is stronger than the ignorance forced upon us. If we realize how indispensable is responsible militant organization to our struggle, we will create it as we managed to create underground railroads, protest groups, self-help societies and the churches that have always been our refuge, our source of hope and our source of action.

Negroes have been slow to organize because they have been traditionally manipulated. The political powers take advantage of three major weaknesses: the manner in which our political leaders emerge; our failure so far to achieve effective political alliances; and the Negro's general reluctance to participate fully in political life.

The majority of Negro political leaders do not ascend to prominence on the shoulders of mass support. Although genuinely popular leaders are now emerging, most are still selected by white leadership, elevated to position, supplied with resources and inevitably subjected to white control. The mass of Negroes nurtures a healthy suspicion toward this manufactured leader, who spends little time in persuading them that he embodies personal integrity, commitment and ability and offers few programs and less service. Tragically, he is in too many respects not a fighter for a new life but a figurehead of the old one. Hence, very few Negro political leaders are impressive or illustrious to their constituents. They enjoy only limited loyalty and qualified support.

This relationship in turn hampers the Negro leader in bargaining with genuine strength and independent firmness with white party leaders. The whites are all too well aware of his impotence and his remoteness from his constituents, and they deal with him as a powerless subordinate. He is accorded a measure of dignity and personal respect but not political power.

The Negro politician therefore finds himself in a vacuum. He has no base in either direction on which to build influence and attain leverage.

In two national polls among Negroes to name their most respected leaders, out of the highest fifteen, only a single political figure, Congressman Adam Clayton Powell, was included and he was in the lower half of both lists. This is in marked contrast to polls in which white people choose their most popular leaders; political personalities are always high on the lists and are represented in goodly numbers. There is no Negro personality evoking affection, respect and emulation to correspond to John F. Kennedy, Eleanor Roosevelt, Herbert Lehman, Earl Warren and Adlai Stevenson, to name but a few.

The circumstances in which Congressman Powell emerged into leadership and the experiences of his career are unique. It would not shed light on the larger picture to attempt to study the very individual factors that apply to him. It is fair to say no other Negro political leader is similar, either in the strengths he possesses, the power he attained or the errors he has committed.

And so we shall have to create leaders who embody virtues we can respect, who have moral and ethical principles we can applaud with an enthusiasm that enables us to rally support for them based on confidence and trust. We will have to demand high standards and give consistent, loyal support to those who merit it. We will have to be a reliable constituency for those who prove themselves to be committed political warriors in our behalf. When our movement has partisan political personalities whose unity with their people is unshakable and whose independence is genuine, they will be treated in white political councils with the respect those who embody such power deserve.

In addition to the development of genuinely independent and representative political leaders, we shall have to master the art of political alliances. Negroes should be natural allies of many white reform and independent political groups, yet they are more commonly organized by old-line machine politicians. We will have to learn to refuse crumbs from the big-city machines and steadfastly demand a fair share of the loaf. When the machine politicians demur, we must be prepared to act in unity and throw our support to such independent parties or reform wings of the major parties as are prepared to take our demands seriously and fight for them vigorously.

The art of alliance politics is more complex and more intricate than it is generally pictured. It is easy to put exciting combinations on paper. It evokes happy memories to recall that our victories in the past decade were won with a broad coalition of organizations representing a wide variety of interests. But we deceive ourselves if we envision the same combination backing structural changes in the society. It did not come together for such a program and will not reassemble for it.

A true alliance is based upon some self-interest of each component group and a common interest into which they merge. For an alliance to have permanence and loyal commitment from its various elements, each of them must have a goal from which it benefits and none must have an outlook in basic conflict with the others.

If we employ the principle of selectivity along these lines, we will find millions of allies who in serving themselves also support us, and on such sound foundations unity and mutual trust and tangible accomplishment will flourish.

In the changing conditions of the South, we will find alliances increasingly instrumental in political progress. For a number of years there were de facto alliances in some states in which Negroes voted for the same candidate as whites because he had shifted from a racist to a moderate position, even though he did not articulate an appeal for Negro votes. In recent years the transformation has accelerated, and many white candidates have entered alliances publicly. As they perceived that the Negro vote was becoming a substantial and permanent factor, they could not remain aloof from it. More and more, competition will develop among white political forces for such a significant bloc of votes, and a monolithic white unity based on racism will no longer be possible.

Racism is a tenacious evil, but it is not immutable. Millions of underprivileged whites are in the process of considering the contradiction between segregation and economic progress. White supremacy can feed their egos but not their stomachs. They will not go hungry or forgo the affluent society to remain racially ascendant.

Governors Wallace and Maddox whose credentials as racists are impeccable, understand this, and for that reason they represent themselves as liberal populists as well. Temporarily they can carry water on both shoulders, but the ground is becoming unsteady beneath their feet. Each of them was faced in the primary last year with a new breed of white southerner who for the first time in history met with Negro organizations to solicit support and championed economic reform without racial demagogy. These new figures won significant numbers of white votes, insufficient for victory but sufficient to point the future directions of the South.

It is true that the Negro vote has not transformed the North; but the fact that northern alliances and political action generally have been poorly executed is no reason to predict that the negative experiences will be automatically extended in the North or duplicated in the South. The northern Negro has never used direct action on a mass scale for reforms, and anyone who predicted ten years ago that the southern Negro would also neglect it would have dramatically been proved in error.

Everything Negroes need will not like magic materialize from the use of the ballot. Yet as a lever of power, if it is given studious attention and employed with the creativity we have proved through our protest activities we possess, it will help to achieve many far-reaching changes during our lifetimes.

The final reason for our dearth of political strength, particularly in the North, arises from the grip of an old tradition on many individual Negroes. They tend to hold themselves aloof from politics as a serious concern. They sense that they are manipulated, and their defense is a cynical disinterest. To safeguard themselves on this front from the exploitation that torments them in so many areas, they shut the door to political activity and retreat into the dark shadows of passivity. Their sense of futility is deep, and in terms of their bitter experiences it is justified. They cannot perceive political action as a source of power. It will take patient and persistent effort to eradicate this mood, but the new consciousness of strength developed in a decade of stirring agitation can be utilized to channel constructive Negro activity into political life and eliminate the stagnation produced by an outdated and defensive paralysis.

In the future we must become intensive political activists. We must be guided in this direction because we need political strength, more desperately than any other group in American society. Most of us are too poor to have adequate economic power, and many of us are too rejected by the culture to be part of any tradition of power. Necessity will draw us toward the power inherent in the creative uses of politics.

Negroes nurture a persisting myth that the Jews of America attained social mobility and status solely because they had money. It is unwise to ignore the error for many reasons. In a negative sense it encourages anti-Semitism and overestimates money as a value. In a positive sense, the full truth reveals a useful lesson.

Jews progressed because they possessed a tradition of education combined with social and political action. The Jewish family enthroned education and sacrificed to get it. The result was far more than abstract learning. Uniting social action with educational competence, Jews became enormously effective in political life. Those Jews who became lawyers, businessmen, writers, entertainers, union leaders and medical men did not vanish into the pursuits of their trade exclusively. They lived an active life in political circles, learning the techniques and arts of politics.

Nor was it only the rich who were involved in social and political action. Millions of Jews for half a century remained relatively poor, but they were far from passive in social and political areas. They lived in homes in which politics was a household word. They were deeply involved in radical parties, liberal parties and conservative parties—they formed many of them. Very few Jews sank into despair and escapism even when discrimination assailed the spirit and corroded initiative. Their life raft in the sea of discouragement was social action.

Without overlooking the towering differences between the Negro and Jewish experiences, the lesson of Jewish mass involvement in social and political action and education is worthy of emulation. Negroes have already started on this road in creating the protest movement, but this is only a beginning. We must involve everyone we can reach, even those with inadequate education, and together acquire political sophistication by discussion, practice and reading.

The many thousands of Negroes who have already found intellectual growth and spiritual fulfillment on this path know its creative possibilities. They are not among the legions of the lost, they are not crushed by the weight of centuries. Most heartening, among the young the spirit of challenge and determination for change is becoming an unquenchable force.

But the scope of struggle is still too narrow and too restricted. We must turn more of our energies and focus our creativity on the useful things that translate into power. We in this generation must do the work and in doing it stimulate our children to learn and acquire higher levels of skill and technique.

It must become a crusade so vital that civil rights organizers do not repeatedly have to make personal calls to summon support. There must be a climate of social pressure in the Negro community that scorns the Negro who will not pick up his citizenship rights and add his strength enthusiastically and voluntarily to the accumulation of power for himself and his people. The past years have blown fresh winds through ghetto stagnation, but we are on the threshold of a significant change that demands a hundredfold acceleration. By 1970 ten of our larger cities will have Negro majorities if present trends continue. We can shrug off this opportunity or use it for a new vitality to deepen and enrich our family and community life.

We must utilize the community action groups and training centers now proliferating in some slum areas to create not merely an electorate, but a conscious, alert and informed people who know their direction and whose collective wisdom and vitality commands respect. The slave heritage can be cast into the dim past by our consciousness of our strengths and a resolute determination to use them in our daily experiences.

Power is not the white man's birthright; it will not be legislated for us and delivered in neat government packages. It is a social force any group can utilize by accumulating its elements in a planned, deliberate campaign to organize it under its own control.

New York Times Magazine (June 11, 1967).

PART IV

A Prophet Foresees the Future

(1967–1968)

18
Where Do We Go from Here?

(1967)

By 1967, Dr. King had become accustomed to being caught between political leftists who accused him of being too cautious, and by those on the political right who accused him of being unpatriotic because of his public criticisms and opposition to racial injustice. Those on the right felt that Dr. King and his cohorts were too eager to embarrass the United States before the eyes of the world, especially since the country was engaged in the "Cold War" struggle against communism. Although the United States and its allies, which included the Union of Soviet Socialist Republics (USSR), defeated the fascist regimes in Germany, Italy and Japan by the end of World War II in 1945, the Western nations were quite uncomfortable with the USSR because it espoused communism. The West feared that the USSR intended to spread the communist doctrine throughout the world.

Mutual fear and hostility spawned a bitter "Cold War"—primarily between the United States and the Soviet Union—that lasted from 1945 to 1987. Under the leadership of Senator Joseph McCarthy of Wisconsin, an era of virulent anticommunism emerged in the U.S. between 1950 and 1954. Public discussions about the nature and destiny of American democracy were hamstrung by McCarthyites who demanded loyalty tests. Civil Rights activists were often accused of being communists, and thus tended to minimize their comments about the Cold War. Dr. King strongly resisted this practice because he was a democratic socialist in his political sentiments, and because he was a preacher who was committed to the Christian gospel of peace, love, and justice. This was Dr. King's last, and most radical, Southern Christian Leadership Conference presidential address, which he delivered at the annual meeting of the SCLC held at Ebenezer Baptist Church in Atlanta, Georgia, on August 16, 1967.

NOW, IN ORDER TO ANSWER THE QUESTION, "WHERE DO WE GO FROM here?" which is our theme, we must first honestly recognize where we are now. When the Constitution was written, a strange formula to determine taxes and representation declared that the Negro was sixty percent of a person. Today another curious formula seems to declare that he is fifty percent of a person. Of the good things in life, the Negro has approximately one half those of whites. Of the bad things of life, he has twice those of whites. Thus half of all Negroes live in substandard housing. And Negroes have half the income of whites. When we view the negative experiences of life, the Negro has a double share. There are twice as many unemployed. The rate of infant mortality among Negroes is double that of whites and there are twice as many Negroes dying in Vietnam as whites in proportion to their size in the population.

In other spheres, the figures are equally alarming. In elementary schools, Negroes lag one to three years behind whites, and their segregated schools receive substantially less money per student than the white schools. One-twentieth as many Negroes as whites attend college. Of employed Negroes, seventy-five percent hold menial jobs.

This is where we are. Where do we go from here? First, we must massively assert our dignity and worth. We must stand up amidst a system that still oppresses us and develop an unassailable and majestic sense of values. We must no longer be ashamed of being black. The job of arousing manhood within a people that have been taught for so many centuries that they are nobody is not easy.

Even semantics have conspired to make that which is black seem ugly and degrading. In Roget's *Thesaurus* there are 120 synonyms for blackness and at least sixty of them are offensive, as for example, blot, soot, grim, devil and foul. And there are some 134 synonyms for whiteness and all are favorable, expressed in such words as purity, cleanliness, chastity and innocence. A white lie is better than a black lie. The most degenerate member of a family is a "black sheep." Ossie Davis has suggested that maybe the English language should be reconstructed so that teachers will not be forced to teach the Negro child sixty ways to despise himself, and thereby perpetuate his false sense of inferiority, and the white child 134 ways to adore himself, and thereby perpetuate his false sense of superiority.

The tendency to ignore the Negro's contribution to American life and to strip him of his personhood is as old as the earliest history books and as contemporary as the morning's newspaper. To upset this cultural homicide, the Negro must rise up with an affirmation of his own Olympian manhood. Any movement for the Negro's freedom that overlooks this necessity is only waiting to be buried. As long as the mind is enslaved, the body can never be free. Psychological freedom, a firm sense of self-esteem, is the most powerful weapon against the long night of physical slavery. No Lincolnian emancipation proclamation or Johnsonian civil rights bill can totally bring this kind of freedom. The Negro will only be free when he reaches down to the inner depths of his own being and signs with the pen and ink of assertive manhood his own emancipation proclamation. And, with a spirit straining toward true self-esteem, the Negro must boldly throw off the manacles of self-abnegation and say to himself and to the world, "I am somebody. I am a person. I am a man with dignity and honor. I have a rich and noble history. How painful and exploited that history has been. Yes, I was a slave through my foreparents and I am not ashamed of that. I'm ashamed of the people who were so sinful to make me a slave." Yes, we must stand up and say, "I'm black and I'm beautiful," and this self-affirmation is the black man's need, made compelling by the white man's crimes against him.

Another basic challenge is to discover how to organize our strength in terms of economic and political power. No one can deny that the Negro is in dire need of this kind of legitimate power. Indeed, one of the great problems that the Negro confronts is his lack of power. From old plantations of the South to newer ghettos of the

North, the Negro has been confined to a life of voicelessness and powerlessness. Stripped of the right to make decisions concerning his life and destiny he has been subject to the authoritarian and sometimes whimsical decisions of this white power structure. The plantation and ghetto were created by those who had power, both to confine those who had no power and to perpetuate their powerlessness. The problem of transforming the ghetto, therefore, is a problem of power—confrontation of the forces of power demanding change and the forces of power dedicated to the preserving of the status quo. Now power properly understood is nothing but the ability to achieve purpose. It is the strength required to bring about social, political and economic change. Walter Reuther defined power one day. He said, "Power is the ability of a labor union like the UAW to make the most powerful corporation in the world, General Motors, say, 'Yes' when it wants to say 'No.' That's power."

Now a lot of us are preachers, and all of us have our moral convictions and concerns, and so often have problems with power. There is nothing wrong with power if power is used correctly. You see, what happened is that some of our philosophers got off base. And one of the great problems of history is that the concepts of love and power have usually been contrasted as opposites—polar opposites—so that love is identified with a resignation of power, and power with a denial of love.

It was this misinterpretation that caused Nietzsche, who was a philosopher of the will to power, to reject the Christian concept of love. It was this same misinterpretation which induced Christian theologians to reject the Nietzschean philosophy of the will to power in the name of the Christian idea of love. Now, we've got to get this thing right. What is needed is a realization that power without love is reckless and abusive, and love without power is sentimental and anemic. Power at its best is love implementing the demands of justice, and justice at its best is power correcting everything that stands against love. And this is what we must see as we move on. What has happened is that we have had it wrong and confused in our own country, and this has led Negro Americans in the past to seek their goals through power devoid of love and conscience.

This is leading a few extremists today to advocate for Negroes the same destructive and conscienceless power that they have justly abhorred in whites. It is precisely this collision of immoral power with powerless morality which constitutes the major crisis of our times.

We must develop a program that will drive the nation to a guaranteed annual income. Now, early in this century this proposal would have been greeted with ridicule and denunciation, as destructive of initiative and responsibility. At that time economic status was considered the measure of the individual's ability and talents. And, in the thinking of that day, the absence of worldly goods indicated a want of industrious habits and moral fiber. We've come a long way in our understanding of human motivation and of the blind operation of our economic system. Now we realize that dislocations in the market operations of our economy and the prevalence of discrimination thrust people into idleness and bind them in constant or frequent unemployment against their will. Today the poor are less often dismissed, I hope, from our consciences by being branded as inferior or incompetent. We also know that no matter how dynamically the economy develops and expands, it does not eliminate all poverty.

The problem indicates that our emphasis must be twofold. We must create full employment or we must create incomes. People must be made consumers by one method or the other. Once they are placed in this position we need to be concerned that the potential of the individual is not wasted. New forms of work that enhance the social good will have to be devised for those for whom traditional jobs are not available. In 1879 Henry George anticipated this state of affairs when he wrote in *Progress and Poverty:*

> "The fact is that the work which improves the condition of mankind, the work which extends knowledge and increases power and enriches literature and elevates thought, is not done to secure a living. It is not the work of slaves driven to their tasks either by the task, by the taskmaster, or by animal necessity. It is the work of men who somehow find a form of work that brings a security for its own sake and a state of society where want is abolished."*

*Henry George (1839–1897) was the father of the single-tax system, which he set forth in his *Progress and Poverty,* published in 1879. The book argued that the land belonged to society, which created its value and properly taxed that value, not improvements on the land.

Work of this sort could be enormously increased, and we are likely to find that the problems of housing and education, instead of preceding the elimination of poverty, will themselves be affected if poverty is first abolished. The poor transformed into purchasers will do a great deal on their own to alter housing decay. Negroes who have a double disability will have a greater effect on discrimination when they have the additional weapon of cash to use in their struggle.

Beyond these advantages, a host of positive psychological changes inevitably will result from widespread economic security. The dignity of the individual will flourish when the decisions concerning his life are in his own hands, when he has the means to seek self-improvement. Personal conflicts among husbands, wives and children will diminish when the unjust measurement of human worth on the scale of dollars is eliminated.

Now our country can do this. John Kenneth Galbraith said that a guaranteed annual income could be done for about twenty billion dollars a year. And I say to you today, that if our nation can spend thirty-five billion dollars a year to fight an unjust, evil war in Vietnam, and twenty billion dollars to put a man on the moon, it can spend billions of dollars to put God's children on their own two feet right here on earth.

Now, let me say briefly that we must reaffirm our commitment to nonviolence. I want to stress this. The futility of violence in the struggle for racial justice has been tragically etched in all the recent Negro riots. Yesterday, I tried to analyze the riots and deal with their causes. Today I want to give the other side. There is certainly something painfully sad about a riot. One sees screaming youngsters and angry adults fighting hopelessly and aimlessly against impossible odds. And deep down within them, you can see a desire for self-destruction, a kind of suicidal longing.

Occasionally Negroes contend that the 1965 Watts riot and the other riots in various cities represented effective civil rights action. But those who express this view always end up with stumbling words when asked what concrete gains have been won as a result. At best, the riots have produced a little additional antipoverty money allotted by frightened government officials, and a few water-sprinklers to cool the children of the ghettos. It is something like improving the food in the prison while the people remain securely incarcerated behind bars. Nowhere have the riots won any concrete improvement such as have the organized protest demonstrations. When one tries to pin down advocates of violence as to what acts would be effective, the answers are blatantly illogical. Sometimes they talk of overthrowing racist state and local governments and they talk about guerrilla warfare. They fail to see that no internal revolution has ever succeeded in overthrowing a government by violence unless the government had already lost the allegiance and effective control of its armed forces. Anyone in his right mind knows that this will not happen in the United States. In a violent racial situation, the power structure has the local police, the state troopers, the National Guard and, finally, the army to call on—all of which are predominantly white. Furthermore, few if any violent revolutions have been successful unless the violent minority had the sympathy and support of the nonresistant majority. Castro may have had only a few Cubans actually fighting with him up in the hills, but he could never have overthrown the Batista regime unless he had the sympathy of the vast majority of Cuban people.*

It is perfectly clear that a violent revolution on the part of American blacks would find no sympathy and support from the white population and very little from the majority of the Negroes themselves. This is no time for romantic illusions and empty philosophical debates about freedom. This is a time for action. What is needed is a strategy for change, a tactical program that will bring the Negro into the mainstream of American life as quickly as possible. So far, this has only been offered by the nonviolent movement. Without recognizing this we will end up with solutions that don't solve, answers that don't answer and explanations that don't explain.

And so I say to you today that I still stand by nonviolence. And I am still convinced that it is the most potent weapon available to the Negro in his struggle for justice in this country. And the other thing is that I am concerned about a better world. I'm concerned about justice. I'm concerned about brotherhood. I'm concerned about truth. And when one is concerned about these, he can never advocate violence. For through violence you may murder a murderer but you

*In 1956 Fidel Castro landed on the coast of Cuba in the vessel, *Gramma,* to overthrow the despot Fulgencio Batista. Twelve men survived the counterattack and went on to lead the Cuban people to victory over Batista, who fled the island on New Year's Day, 1959, which ushered in the Cuban revolutionary victory.

can't murder murder. Through violence you may murder a liar but you can't establish truth. Through violence you may murder a hater, but you can't murder hate. Darkness cannot put out darkness. Only light can do that.

And I say to you, I have also decided to stick to love. For I know that love is ultimately the only answer to mankind's problems. And I'm going to talk about it everywhere I go. I know it isn't popular to talk about it in some circles today. I'm not talking about emotional bosh when I talk about love, I'm talking about a strong, demanding love. And I have seen too much hate. I've seen too much hate on the faces of sheriffs in the South. I've seen hate on the faces of too many Klansmen and too many White Citizens Councilors in the South to want to hate myself, because every time I see it, I know that it does something to their faces and their personalities and I say to myself that hate is too great a burden to bear. I have decided to love. If you are seeking the highest good, I think you can find it through love. And the beautiful thing is that we are moving against wrong when we do it, because John was right, God is love. He who hates does not know God, but he who has love has the key that unlocks the door to the meaning of ultimate reality.

I want to say to you as I move to my conclusion, as we talk about "Where do we go from here," that we honestly face the fact that the movement must address itself to the question of restructuring the whole of American society. There are forty million poor people here. And one day we must ask the question, "Why are there forty million poor people in America?" And when you begin to ask that question, you are raising questions about the economic system, about a broader distribution of wealth. When you ask that question, you begin to question the capitalistic economy. And I'm simply saying that more and more, we've got to begin to ask questions about the whole society. We are called upon to help the discouraged beggars in life's marketplace. But one day we must come to see that an edifice which produces beggars needs restructuring. It means that questions must be raised. You see, my friends, when you deal with this, you begin to ask the question, "Who owns the oil?" You begin to ask the question, "Who owns the iron ore?" You begin to ask the question, "Why is it that people have to pay water bills in a world that is two-thirds water?" These are questions that must be asked.

Now, don't think that you have me in a "bind" today. I'm not talking about communism.

What I'm saying to you this morning is that communism forgets that life is individual. Capitalism forgets that life is social, and the kingdom of brotherhood is found neither in the thesis of communism nor the antithesis of capitalism but in a higher synthesis. It is found in a higher synthesis that combines the truths of both. Now, when I say question the whole society, it means ultimately coming to see that the problem of racism, the problem of economic exploitation, and the problem of war are all tied together. These are the triple evils that are interrelated.

If you will let me be a preacher just a little bit—One night, a juror came to Jesus and he wanted to know what he could do to be saved. Jesus didn't get bogged down in the kind of isolated approach of what he shouldn't do. Jesus didn't say, "Now Nicodemus, you must stop lying." He didn't say, "Nicodemus, you must stop cheating if you are doing that." He didn't say, "Nicodemus, you must not commit adultery." He didn't say, "Nicodemus, now you must stop drinking liquor if you are doing that excessively." He said something altogether different, because Jesus realized something basic—that if a man will lie, he will steal. And if a man will steal, he will kill. So instead of just getting bogged down in one thing, Jesus looked at him and said, "Nicodemus, you must be born again."

He said, in other words, "Your whole structure must be changed." A nation that will keep people in slavery for 244 years will "thingify" them—make them things. Therefore they will exploit them, and poor people generally, economically. And a nation that will exploit economically will have to have foreign investments and everything else, and will have to use its military might to protect them. All of these problems are tied together. What I am saying today is that we must go from this convention and say, "America, you must be born again!"

So, I conclude by saying again today that we have a task and let us go out with a "divine dissatisfaction." Let us be dissatisfied until America will no longer have a high blood pressure of creeds and an anemia of deeds. Let us be dissatisfied until the tragic walls that separate the outer city of wealth and comfort and the inner city of poverty and despair shall be crushed by the battering rams of the forces of justice. Let us be dissatisfied until those that live on the outskirts of hope are brought into the metropolis of daily security. Let us be dissatisfied until slums are cast into the junk heaps of history, and every family is living in a decent sanitary home. Let us be dissatisfied until the dark yesterdays of segregated schools will be transformed into bright tomorrows of quality, integrated education. Let us be dissatisfied until integration is not seen as a problem but as an opportunity to participate in the beauty of diversity. Let us be dissatisfied until men and women, however black they may be, will be judged on the basis of the content of their character and not on the basis of the color of their skin. Let us be dissatisfied. Let us be dissatisfied until every state capitol houses a governor who will do justly, who will love mercy and who will walk humbly with his God. Let us be dissatisfied until from every city hall, justice will roll down like waters and righteousness like a mighty stream. Let us be dissatisfied until that day when the lion and the lamb shall lie down together, and every man will sit under his own vine and fig tree and none shall be afraid. Let us be dissatisfied. And men will recognize that out of one blood God made all men to dwell upon the face of the earth. Let us be dissatisfied until that day when nobody will shout "White Power!"—when nobody will shout "Black Power!"—but everybody will talk about God's power and human power.

I must confess, my friends, the road ahead will not always be smooth. There will be still rocky places of frustration and meandering points of bewilderment. There will be inevitable setbacks here and there. There will be those moments when the buoyancy of hope will be transformed into the fatigue of despair. Our dreams will sometimes be shattered and our ethereal hopes blasted. We may again with tear-drenched eyes have to stand before the bier of some courageous civil rights worker whose life will be snuffed out by the dastardly acts of bloodthirsty mobs. Difficult and painful as it is, we must walk on in the days ahead with an audacious faith in the future. And as we continue our charted course, we may gain consolation in the words so nobly left by that great black bard who was also a great freedom fighter of yesterday, James Weldon Johnson:

> Stony the road we trod,
> Bitter the chastening rod
> Felt in the days
> When hope unborn had died.

Yet with a steady beat,
Have not our weary feet
Come to the place
For which our fathers sighed?

We have come over the way
That with tears hath been watered.
We have come treading our paths
Through the blood of the slaughtered,

Out from the gloomy past,
Till now we stand at last
Where the bright gleam
Of our bright star is cast.

Let this affirmation be our ringing cry. It will give us the courage to face the uncertainties of the future. It will give our tired feet new strength as we continue our forward stride toward the city of freedom. When our days become dreary with low-hovering clouds of despair, and when our nights become darker than a thousand midnights, let us remember that there is a creative force in this universe, working to pull down the gigantic mountains of evil, a power that is able to make a way out of no way and transform dark yesterdays into bright tomorrows. Let us realize the arc of the moral universe is long but it bends toward justice.

Let us realize that William Cullen Bryant is right: "Truth crushed to earth will rise again." Let us go out realizing that the Bible is right: "Be not deceived, God is not mocked. Whatsoever a man soweth, that shall he also reap." This is for hope for the future, and with this faith we will be able to sing in some not too distant tomorrow with a cosmic past tense, "We have overcome, we have overcome, deep in my heart, I did believe we would overcome."

This speech was published under the title "New Sense of Direction" in *Worldview* 15 (April 1972): 5ff.

19

The Drum Major Instinct

(1968)

Black churches were the religious parent of African American culture. As stated in the introduction, these congregations became visible in the late eighteenth century and deeply influenced virtually every movement to better the lives of African American people throughout the course of their history in this country. Ebenezer Baptist Church in Atlanta, Georgia, the congregation that nurtured and ordained Dr. King, aptly illustrates the significance of the Black Church as a center of religious and ethical values, as well as a center of social and political activism. The early Ebenezer was formed in 1886 partly as an outgrowth of the older Wheat Street Baptist Church. It did not thrive until the Reverend A. D. Williams, the maternal grandfather of Martin Luther King, Sr., became its pastor in 1894, just one year before the founding of the National Baptist Convention. This convention was, and still is, the largest organization of African American people in the United States.

Under the leadership of Dr. King's grandfather, Ebenezer immediately became a part of this conclave of significant churches and preachers who later provided the basic national, spiritual and material support that sustained the Southern Christian Leadership Conference. After Dr. King's grandfather died in 1931, his father, the Reverend Martin Luther King, Sr., often called "Daddy King," was elected by the Ebenezer congregation to succeed his father-in-law as the pastor of Ebenezer. He served until his death in 1975. Dr. King's father became an important partner in the amazing informal consortium of African American enterprises and churches that were located on Auburn Avenue in Atlanta. They included the Atlanta Life Insurance Company, the Prince Hall Masonic Lodge (where the offices of the Southern Christian Leadership Conference were later housed), Wheat Street Baptist Church, and Ebenezer Baptist Church.

The Kings lived on Auburn Avenue, just a few houses from Ebenezer. Martin Luther King, Jr., went to Sunday school at Ebenezer, formed some of his closest and most enduring friendships there and in the Auburn Avenue neighborhood. He felt comfortable and loved by the people of Ebenezer whose voluntary offerings of money, time and talent helped to make the ministry of the King family possible. Given these factors, it is understandable that Dr. King gave his most poignant sermons and addresses from the pulpit of Ebenezer Baptist Church where he served as co-pastor with his father from 1960 to his assassination in 1968. Dr. King preached the following prophetic and highly personal sermon from the pulpit of Ebenezer Baptist Church on February 4, 1968. Excerpts from it were played at his nationally televised funeral service held at Ebenezer Baptist Church on April 9, 1968, five days after his assassination. Two months before he was murdered at the age of 39, Dr. King somehow sensed that he would not live a long life.

THIS MORNING I WOULD LIKE TO USE AS A SUBJECT FROM WHICH TO preach "The Drum Major Instinct." And our text for the morning is taken from a very familiar passage in the tenth chapter as recorded by Saint Mark; beginning with the thirty-fifth verse of that chapter, we read these words: "And James and John the sons of Zebedee came unto him saying, 'Master, we would that thou shouldest do for us whatsoever we shall desire.' And he said unto them, 'What would ye that I should do for you?' And they said unto him, 'Grant unto us that we may sit one on thy right hand, and the other on thy left hand in thy glory.' But Jesus said unto them, 'Ye know not what ye ask. Can ye drink of the cup that I drink of, and be baptized with the baptism that I am baptized with?' And they said unto him, 'We can.' And Jesus said unto them, 'Ye shall indeed drink of the cup that I drink of, and with the baptism that I am baptized with all shall ye be baptized. But to sit on my right hand and on my left hand is not mine to give, but it shall be given to them for whom it is prepared.' "

And then, Jesus goes on toward the end of that passage to say, "But so shall it not be among you, but whosoever will be great among you, shall be your servant; and whosoever of you will be the chiefest, shall be servant of all." The setting is clear. James and John are making a specific request of the master. They had dreamed, as most Hebrews dreamed, of a coming king of Israel who would set Jerusalem free. And establish his kingdom on Mount Zion, and in righteousness rule the world. And they thought of Jesus as this kind of king, and they were thinking of that day when Jesus would reign supreme as this new king of Israel. And they were saying now, "when you establish your kingdom, let one of us sit on the right hand, and the other on the left hand of your throne."

Now very quickly, we would automatically condemn James and John, and we would say they were selfish. Why would they make such a selfish request? But before we condemn them too quickly, let us look calmly and honestly at ourselves, and we will discover that we too have those same basic desires for recognition, for importance, that same desire for attention, that same desire to be first. Of course the other disciples got mad with James and John, and you could understand why, but we must understand that we have some of the same James and John qualities. And there is, deep down within all of us, an instinct. It's a kind of drum major instinct—a desire to be out front, a desire to lead the parade, a desire to be first. And it is something that runs a whole gamut of life.

And so before we condemn them, let us see that we all have the drum major instinct. We all want to be important, to surpass others, to achieve distinction, to lead the parade. Alfred Adler, the great psychoanalyst, contends that this is the dominant impulse. Sigmund Freud used to contend that sex was the dominant impulse, and Adler came with a new argument saying that this quest for recognition, this desire for attention, this desire for distinction is the basic impulse, the basic drive of human life—this drum major instinct.

And you know, we begin early to ask life to put us first. Our first cry as a baby was a bid for attention. And all through childhood the drum major impulse or instinct is a major obsession. Children ask life to grant them first place. They are a little bundle of ego. And they have innately the drum major impulse, or the drum major instinct.

Now in adult life, we still have it, and we really never get by it. We like to do something good. And you know, we like to be praised for it. Now if you don't believe that, you just go on living life, and you will discover very soon that you like to be praised. Everybody likes it, as a matter of fact. And somehow this warm glow we feel when we are praised, or when our name is in print, is something of the vitamin A to our ego. Nobody is unhappy when they are praised, even if they know they don't deserve it, and even if they don't believe it. The only unhappy people about praise is when that praise is going too much toward somebody else. But everybody likes to be praised, because of this real drum major instinct.

Now the presence of the drum major instinct is why so many people are joiners. You know there are some people who just join everything. And it's really a quest for attention, and recognition, and importance. And they get names that give them that impression. So you get your groups, and they become the grand patron, and the little fellow who is henpecked at home needs a chance to be the most worthy of the most worthy of something. It is the drum major impulse and longing that runs the gamut of human life. And so we see it everywhere, this quest for recognition. And we join things, over-join really, that we think that we will find that recognition in.

Now the presence of this instinct explains why we are so often taken by advertisers. You know those gentlemen of massive verbal persuasion. And they have a way of saying things to you that kind of gets you into buying. In order to be a man of distinction, you must drink this whiskey. In order to make your neighbors envious, you must drive this type of car. In order to be lovely to love you must wear this kind of lipstick or this kind of perfume. And you know, before you know it you're just buying that stuff. That's the way the advertisers do it.

I got a letter the other day. It was a new magazine coming out. And it opened up, "Dear Dr. King. As you know, you are on many mailing lists. And you are categorized as highly intelligent, progressive, a lover of the arts, and the sciences, and I know you will want to read what I have to say." Of course I did. After you said all of that and explained me so exactly, of course I wanted to read it.

But very seriously, it goes through life, the drum major instinct is real. And you know what else it causes to happen? It often causes us to live above our means. It's nothing but the drum major instinct. Do you ever see people buy cars that they can't even begin to buy in terms of their income? You've seen people riding around in Cadillacs and Chryslers who don't earn enough to have a good Model-T Ford. But it feeds a repressed ego.

You know economists tell us that your automobiles should not cost more than half of your annual income. So if you're making an income of five thousand dollars, your car shouldn't cost more than about twenty-five hundred. That's just good economics. And if it's a family of two, and both members of the family make ten thousand dollars, they would have to make out with one car. That would be good economics, although it's often inconvenient. But so often . . . haven't you seen people making five thousand dollars a year and driving a car that costs six thousand. And they wonder why their ends never meet. That's a fact.

Now the economists also say that your house shouldn't cost, if you're buying a house, it shouldn't cost more than twice your income. That's based on the economy, and how you would make ends meet. So, if you have an income of five thousand dollars, it's kind of difficult in this society. But say it's a family with an income of ten thousand dollars, the house shouldn't cost more than twenty thousand. But I've seen folk making ten thousand dollars, living in a forty- and fifty-thousand-dollar house. And you know they just barely make it. They get a check every month somewhere, and they owe all of that out before it comes in; never have anything to put away for rainy days.

But now the problem is, it is the drum major instinct. And you know, you see people over and over again with the drum major instinct taking them over. And they just live their lives trying to outdo the Joneses. They got to get this coat because this particular coat is a little better, and a little better-looking than Mary's coat. And I got to drive this car because it's something about this car that makes my car a little better than my neighbor's car. I know a man who used to live in a thirty-five-thousand-dollar house. And other people started building thirty-five-thousand-dollar houses, so he built a seventy-thousand-dollar house, and he built a hundred-thousand-dollar house. And I don't know where he's going to end up if he's going to live his life trying to keep up with the Joneses.

There comes a time that the drum major instinct can become destructive. And that's where I want to move now. I want to move to the point of saying that if this instinct is not harnessed, it becomes a very dangerous, pernicious instinct. For instance, if it isn't harnessed, it causes one's personality to become distorted. I guess that's the most damaging aspect of it—what it does to the personality. If it isn't harnessed, you will end up day in and day out trying to deal with your ego problem by boasting.

Have you ever heard people that—you know, and I'm sure you've met them—that really become sickening because they just sit up all the time talking about themselves. And they just boast, and boast, and boast, and that's the person who has not harnessed the drum major instinct.

And then it does other things to the personality. It causes you to lie about who you know sometimes. There are some people who are influence peddlers. And in their attempt to deal with the drum major instinct, they have to try to identify with the so-called big name people. And if you're not careful, they will make you think they know somebody that they don't really know. They know them well, they sip tea with them. And they . . . this and that. That . . . that happens to people.

And the other thing is that it causes one to engage ultimately in activities that are merely used to get attention. Criminologists tell us that some people are driven to crime because of this drum major instinct. They don't feel that they are getting enough attention through the normal channels of social behavior, and others turn to anti-social behavior in order to get attention, in order to feel important. And so they get that gun. And before they know it they rob the bank in a quest for recognition, in a quest for importance.

And then the final great tragedy of the distorted personality is the fact that when one fails to harness this instinct, he ends by trying to push others down in order to push himself up. And whenever you do that, you engage in some of the most vicious activities. You will spread evil, vicious, lying gossip on people, because you are trying to pull them down in order to push yourself up.

And the great issue of life is to harness the drum major instinct.

Now the other problem is when you don't harness the drum major instinct, this uncontrolled aspect of it, is that it leads to snobbish exclusivism. Now you know, this is the danger of social clubs, and fraternities. I'm in a fraternity; I'm in two or three. For sororities, and all of these, I'm not talking against them, I'm saying it's the danger. The danger is that they can become forces of classism and exclusivism where somehow you get a degree of satisfaction because you are in something exclusive, and that's fulfilling something, you know. And I'm in this fraternity, and it's the best fraternity in the world and everybody can't get in this fraternity. So it ends up, you know, a very exclusive kind of thing.

And you know, that can happen with the church. I've known churches get in that bind sometimes. I've been to churches you know, and they say, "We have so many doctors and so many schoolteachers, and so many lawyers, and so many businessmen in our church." And that's fine, because doctors need to go to church, and lawyers, and businessmen, teachers—they ought to be in church. But they say that, even the preacher sometimes will go on through it, they say that as if the other people don't count. And the church is the one place where a doctor ought to forget that he's a doctor. The church is the one place where a Ph.D. ought to forget that he's a Ph.D. The church is the one place that a schoolteacher ought to forget the degree she has behind her name. The church is the one place where the lawyer ought to forget that he's a lawyer. And any church that violates the 'whosoever will, let him come' doctrine is a dead, cold church, and nothing but a little social club with a thin veneer of religiosity.

When the church is true to its nature, it says, "Whosoever will, let him come." And it does not propose to satisfy the perverted uses of the drum major instinct. It's the one place where everybody should be the same standing before a common master and savior. And a recognition grows out of this—that all men are brothers because they are children of a common father.

The drum major instinct can lead to exclusivism in one's thinking, and can lead one to feel that because he has some training, he's a little better than that person that doesn't have it, or because he has some economic security, that he's a little better than the person who doesn't have it. And that's the uncontrolled, perverted use of the drum major instinct.

Now the other thing is that it leads to tragic—and we've seen it happen so often—tragic race prejudice. Many have written about this problem—Lillian Smith used to say it beautifully in some of her books. And she would say it to the point of getting men and women to see the source of the problem. Do you know that a lot of the race problem grows out of the drum major instinct? A need that some people have to feel superior. A need that some people have to feel that they are first, and to feel that their white skin ordained them to be first. And they have said it over and over again in ways that we see with our own eyes. In fact, not too long ago, a man down in Mississippi said that God was a charter member of the White Citizens Council. And so God being the charter member means that everybody who's in that has a kind of divinity, a kind of superiority.

And think of what has happened in history as a result of this perverted use of the drum major instinct. It has led to the most tragic prejudice, the most tragic expressions of man's inhumanity to man.

I always try to do a little converting when I'm in jail. And when we were in jail in Birmingham the other day, the white wardens all enjoyed coming around to the cell to talk about the race problem. And they were showing us where we were so wrong demonstrating. And they were showing us where segregation was so right. And they were showing us where intermarriage was so wrong. So I would get to preaching, and we would get to talking—calmly, because they wanted to talk about it. And then we got down one day to the point—that was the second or third day—to talk about where they lived, and how much they were earning. And when those brothers told me what they were earning, I said, now "You know what? You ought to be marching with us. You're just as poor as Negroes." And I said, "You are put in the position of supporting your oppressor. Because through prejudice and blindness, you fail to see that the same forces that oppress Negroes in American society oppress poor white people. And all you are living on is the satisfaction of your skin being white, and the drum major instinct of thinking that you are somebody big because you are white. And you're so poor you can't send your children to school. You ought to be out here marching with every one of us every time we have a march."

Now that's a fact. That the poor white has been put into this position—where through blindness and prejudice, he is forced to support his oppressors, and the only thing he has going for him is the false feeling that he is superior because his skin is white. And can't hardly eat and make his ends meet week in and week out.

And not only does this thing go into the racial struggle, it goes into the struggle between nations. And I would submit to you this morning that what is wrong in the world today is that the nations of the world are engaged in a bitter, colossal contest for supremacy. And if something doesn't happen to stop this trend I'm sorely afraid that we won't be here to talk about Jesus Christ and about God and about brotherhood too many more years. If somebody doesn't bring an end to this suicidal thrust that we see in the world today, none of us are going to be around, because somebody's going to make the mistake through our senseless blundering of dropping a nuclear bomb somewhere, and then another one is going to drop. And don't let anybody fool you, this can happen within a matter of seconds. They have twenty-megaton bombs in Russia right now that can destroy a city as big as New York in three seconds with everybody wiped away, and every building. And we can do the same thing to Russia and China.

But this is where we are drifting, and we are drifting there, because nations are caught up with the drum major instinct. I must be first. I must be supreme. Our nation must rule the world. And I am sad to say that the nation in which we live is the supreme culprit. And I'm going to continue to say it to America, because I love this country too much to see the drift that it has taken.

God didn't call America to do what she's doing in the world now. God didn't call America to engage in a senseless, unjust war, [such] as the war in Vietnam. And we are criminals in that war. We have committed more war crimes almost than any nation in the world, and I'm going to continue to say it. And we won't stop it because of our pride, and our arrogance as a nation.

But God has a way of even putting nations in their place. The God that I worship has a way of saying, "Don't play with me." He has a way of saying, as the God of the Old Testament used to say to the

Hebrews, "Don't play with me, Israel. Don't play with me, Babylon. Be still and know that I'm God. And if you don't stop your reckless course, I'll rise up and break the backbone of your power." And that can happen to America. Every now and then I go back and read Gibbons' *Decline and Fall of the Roman Empire*. And when I come and look at America, I say to myself, the parallels are frightening.

And we have perverted the drum major instinct. But let me rush on to my conclusion, because I want you to see what Jesus was really saying. What was the answer that Jesus gave these men? It's very interesting. One would have thought that Jesus would have said, "You are out of your place. You are selfish. Why would you raise such a question?"

But that isn't what Jesus did. He did something altogether different. He said in substance, "Oh, I see, you want to be first. You want to be great. You want to be important. You want to be significant. Well you ought to be. If you're going to be my disciple, you must be." But he reordered priorities. And he said, "Yes, don't give up this instinct. It's a good instinct if you use it right. It's a good instinct if you don't distort it and pervert it. Don't give it up. Keep feeling the need for being important. Keep feeling the need for being first. But I want you to be first in love. I want you to be first in moral excellence. I want you to be first in generosity. That is what I want you to do."

And he transformed the situation by giving a new definition of greatness. And you know how he said it? He said now, "Brethren, I can't give you greatness. And really, I can't make you first." This is what Jesus said to James and John. You must earn it. True greatness comes not by favoritism, but by fitness. And the right hand and the left are not mine to give, they belong to those who are prepared.

And so Jesus gave us a new norm of greatness. If you want to be important—wonderful. If you want to be recognized—wonderful. If you want to be great—wonderful. But recognize that he who is greatest among you shall be your servant. That's your new definition of greatness. And this morning, the thing that I like about it . . . by giving that definition of greatness, it means that everybody can be great. Because everybody can serve. You don't have to have a college degree to serve. You don't have to make your subject and your verb agree to serve. You don't have to know about Plato and Aristotle to serve. You don't have to know Einstein's theory of relativity to serve. You don't have to know the second theory of thermodynamics in physics to serve. You only need a heart full of grace. A soul generated by love. And you can be that servant.

I know a man, and I just want to talk about him a minute, and maybe you will discover who I'm talking about as I go down the way, because he was a great one. And he just went about serving. He was born in an obscure village, the child of a poor peasant woman. And then he grew up in still another obscure village, where he worked as a carpenter until he was thirty years old. Then for three years, he just got on his feet, and he was an itinerant preacher. And then he went about doing some things. He didn't have much. He never wrote a book. He never held an office. He never had a family. He never owned a house. He never went to college. He never visited a big city. He never went two hundred miles from where he was born. He did none of the usual things that the world would associate with greatness. He had no credentials but himself.

He was thirty-three when the tide of public opinion turned against him. They called him a rabble-rouser. They called him a troublemaker. They said he was an agitator. He practiced civil disobedience; he broke injunctions. And so he was turned over to his enemies, and went through the mockery of a trial. And the irony of it all is that his friends turned him over to them. One of his closest friends denied him. Another of his friends turned him over to his enemies. And while he was dying, the people who killed him gambled for his clothing, the only possession that he had in the world. When he was dead, he was buried in a borrowed tomb, through the pity of a friend.

Nineteen centuries have come and gone, and today, he stands as the most influential figure that ever entered human history. All of the armies that ever marched, all the navies that ever sailed, all the parliaments that ever sat, and all the kings that ever reigned put together have not affected the life of man on this earth as much as that one solitary life. His name may be a familiar one. But today I can hear them talking about him. Every now and then somebody says, "He's king of kings." And again I can hear somebody saying, "He's lord of lords." Somewhere else I can hear somebody saying, "In Christ there is no east nor west." And they go on and talk about. . . . "In him there's no north and south, but one great fellowship of love throughout the whole wide world." He didn't have anything. He just went around serving, and doing good.

This morning, you can be on his right hand and his left hand if you serve. It's the only way in.

Every now and then I guess we all think realistically about that day when we will be victimized with what is life's final common denominator—that something we call death. We all think about it. And every now and then I think about my own death, and I think about my own funeral. And I don't think of it in a morbid sense. Every now and then I ask myself, "What is it that I would want said?" And I leave the word to you this morning.

If any of you are around when I have to meet my day, I don't want a long funeral. And if you get somebody to deliver the eulogy, tell them not to talk too long. Every now and then I wonder what I want them to say. Tell them not to mention that I have a Nobel Peace Prize, that isn't important. Tell them not to mention that I have three or four hundred other awards, that's not important. Tell him not to mention where I went to school.

I'd like somebody to mention that day, that Martin Luther King, Jr., tried to give his life serving others. I'd like for somebody to say that day, that Martin Luther King, Jr., tried to love somebody. I want you to say that day, that I tried to be right on the war question. I want you to be able to say that day, that I did try to feed the hungry. And I want you to be able to say that day, that I did try, in my life, to clothe those who were naked. I want you to say, on that day, that I did try, in my life, to visit those who were in prison. I want you to say that I tried to love and serve humanity.

Yes, if you want to say that I was a drum major, say that I was a drum major for justice; say that I was a drum major for peace; I was a drum major for righteousness. And all of the other shallow things will not matter. I won't have any money to leave behind. I won't have the fine and luxurious things of life to leave behind. But I just want to leave a committed life behind.

And that's all I want to say . . . if I can help somebody as I pass along, if I can cheer somebody with a word or song, if I can show somebody he's traveling wrong, then my living will not be in vain. If

I can do my duty as a Christian ought, if I can bring salvation to a world once wrought, if I can spread the message as the master taught, then my living will not be in vain.

Yes, Jesus, I want to be on your right side or your left side, not for any selfish reason. I want to be on your right or your best side, not in terms of some political kingdom or ambition, but I just want to be there in love and in justice and in truth and in commitment to others, so that we can make of this old world a new world.

Flip Schulke, ed. *Martin Luther King, Jr.: A Documentary . . . Montgomery to Memphis* (New York and London: Norton, 1976), 220–22.

20

I See the Promised Land

(1968)

Martin Luther King, Jr., had a deep sense of religious vocation. He believed that somehow God had called him to encourage the United States of America to admit that it had mistreated its citizens of African descent who had known slavery, racial segregation, economic discrimination, and deprivation. But the cost of this high sense of personal purpose was to earn the enmity of his fellow citizens who did not believe that it was right for a minister to lead social and political crusades. Dr. King fervently believed religious ministers have public responsibilities, that the war in Vietnam was immoral, and that injustice in any form is not to be tolerated in the name of securing a sense of cheap peace and security. This is why he answered the call of the predominantly African American sanitation workers in Memphis, Tennessee, to come to their city in order to help them in their strike against unfair labor practices.

Just a few days before he gave the following speech, one of his demonstrations was taken over by disgruntled black people who had grown weary of the slow pace of Dr. King's nonviolent resistance strategy. Looting, many injuries and shootings ensued. Dr. King confessed privately that he had become depressed. He felt like a failure. But there were many people who still believed in him and in the power of nonviolent resistance. They rallied to offer him, and the movement, encouragement and hope. Of course Dr. King was the principal speaker. It was a poignant moment. It was his last sermon and one of his greatest speeches. He delivered it on April 3, 1968, on what turned out to be the eve of his assassination, at Mason Temple, a church in Memphis.

THANK YOU VERY KINDLY, MY FRIENDS. AS I LISTENED TO RALPH ABERNATHY in his eloquent and generous introduction and then thought about myself, I wondered who he was talking about. It's always good to have your closest friend and associate say something good about you. And Ralph is the best friend that I have in the world.

I'm delighted to see each of you here tonight in spite of a storm warning. You reveal that you are determined to go on anyhow. Something is happening in Memphis, something is happening in our world.

As you know, if I were standing at the beginning of time, with the possibility of general and panoramic view of the whole human history up to now, and the Almighty said to me, "Martin Luther King, which age would you like to live in?"—I would take my mental flight by Egypt through, or rather across the Red Sea, through the wilderness on toward the promised land. And in spite of its magnificence, I wouldn't stop there. I would move on by Greece, and take my mind to Mount Olympus. And I would see Plato, Aristotle, Socrates, Euripides and Aristophanes assembled around the Parthenon as they discussed the great and eternal issues of reality.

But I wouldn't stop there. I would go on, even to the great heyday of the Roman Empire. And I would see developments around there, through various emperors and leaders. But I wouldn't stop there. I would even come up to the day of the Renaissance, and get a quick picture of all that the Renaissance did for the cultural and esthetic life of man. But I wouldn't stop there. I would even go by the way that the man for whom I'm named had his habitat. And I would watch Martin Luther as he tacked his ninety-five theses on the door at the church in Wittenberg.

But I wouldn't stop there. I would come on up even to 1863, and watch a vacillating president by the name of Abraham Lincoln finally come to the conclusion that he had to sign the Emancipation Proclamation. But I wouldn't stop there, I would even come up to the early thirties, and see a man grappling with the problems of the bankruptcy of his nation. And come with an eloquent cry that we have nothing to fear but fear itself.

But I wouldn't stop there. Strangely enough, I would turn to the Almighty, and say, "If you allow me to live just a few years in the second half of the twentieth century, I will be happy." Now that's a strange statement to make, because the world is all messed up. The nation is sick. Trouble is in the land. Confusion all around. That's a strange statement. But I know, somehow, that only when it is dark enough, can you see the stars. And I see God working in this period of the twentieth century in a way that men, in some strange way, are responding—something is happening in our world. The masses of people are rising up. And wherever they are assembled today, whether they are in Johannesburg, South Africa; Nairobi, Kenya; Accra, Ghana; New York City; Atlanta, Georgia; Jackson, Mississippi; or Memphis, Tennessee—the cry is always the same—"We want to be free."

And another reason that I'm happy to live in this period is that we have been forced to a point where we're going to have to grapple with the problems that men have been trying to grapple with through history, but the demands didn't force them to do it. Survival demands that we grapple with them. Men, for years now, have been talking about war and peace. But now, no longer can they just talk about it. It is no longer a choice between violence and nonviolence in this world; it's nonviolence or nonexistence.

That is where we are today. And also in the human rights revolution, if something isn't done, and in a hurry, to bring the colored peoples of the world out of their long years of poverty, their long years of hurt and neglect, the whole world is doomed. Now, I'm just happy that God has allowed me to live in this period, to see what is unfolding. And I'm happy that he's allowed me to be in Memphis.

I can remember, I can remember when Negroes were just going around as Ralph has said, so often, scratching where they didn't itch, and laughing when they were not tickled. But that day is all over. We mean business now, and we are determined to gain our rightful place in God's world.

And that's all this whole thing is about. We aren't engaged in any negative protest and in any negative arguments with anybody. We are saying that we are determined to be men. We are determined to be people. We are saying that we are God's children. And that we don't have to live like we are forced to live.

Now, what does all of this mean in this great period of history? It means that we've got to stay together. We've got to stay together and maintain unity. You know, whenever Pharaoh wanted to prolong the period of slavery in Egypt, he had a favorite, favorite formula for doing it. What was that? He kept the slaves fighting among themselves. But whenever the slaves get together, something happens in Pharaoh's court, and he cannot hold the slaves in slavery. When the slaves get together, that's the beginning of getting out of slavery. Now let us maintain unity.

Secondly, let us keep the issues where they are. The issue is injustice. The issue is the refusal of Memphis to be fair and honest in its dealings with its public servants, who happen to be sanitation workers. Now, we've got to keep attention on that. That's always the problem with a little violence. You know what happened the other day, and the press dealt only with the window-breaking. I read the articles. They very seldom got around to mentioning the fact that one thousand, three hundred sanitation workers were on strike, and that Memphis is not being fair to them, and that Mayor Loeb is in dire need of a doctor. They didn't get around to that.

Now we're going to march again, and we've got to march again, in order to put the issue where it is supposed to be. And force everybody to see that there are thirteen hundred of God's children here suffering, sometimes going hungry, going through dark and dreary nights wondering how this thing is going to come out. That's the issue. And we've got to say to the nation: we know it's coming out. For when people get caught up with that which is right and they are willing to sacrifice for it, there is no stopping point short of victory.

We aren't going to let any mace stop us. We are masters in our nonviolent movement in disarming police forces; they don't know what to do. I've seen them so often. I remember in Birmingham, Alabama, when we were in that majestic struggle there we would move out of the 16th Street Baptist Church day after day; by the hundreds we would move out. And Bull Connor would tell them to send the dogs forth and they did come; but we just went before the dogs singing, "Ain't gonna let nobody turn me round." Bull Connor next would say, "Turn the fire hoses on." And as I said to you the other night, Bull Connor didn't know history. He knew a kind of physics that somehow didn't relate to the transphysics that we knew about. And that was the fact that there was a certain kind of fire that no water could put out. And we went before the fire hoses; we had known water. If we were Baptist or some other denomination, we had been immersed. If we were Methodist, and some others, we had been sprinkled, but we knew water.

That couldn't stop us. And we just went on before the dogs and we would look at them; and we'd go on before the water hoses and we would look at it, and we'd just go on singing "Over my head I see freedom in the air." And then we would be thrown in the paddy wagons, and sometimes we were stacked in there like sardines in a can. And they would throw us in, and old Bull would say, "Take them off," and they did; and we would just go in the paddy wagon singing, "We Shall Overcome." And every now and then we'd get in the jail, and we'd see the jailers looking through the windows being moved by our prayers, and being moved by our words and our songs. And there was a power there which Bull Connor couldn't adjust to; and so we ended up transforming Bull into a steer, and we won our struggle in Birmingham.

Now we've got to go on to Memphis just like that. I call upon you to be with us Monday. Now about injunctions: We have an injunction and we're going into court tomorrow morning to fight this illegal, unconstitutional injunction. All we say to America is, "Be true to what you said on paper." If I lived in China or even Russia, or any totalitarian country, maybe I could understand the denial of certain basic First Amendment privileges, because they hadn't committed themselves to that over there. But somewhere I read of the freedom of assembly. Somewhere I read of the freedom of speech. Somewhere I read of the freedom of the press. Somewhere I read that the greatness of America is the right to protest for right. And so just as I say, we aren't going to let any injunction turn us around. We are going on.

We need all of you. And you know what's beautiful to me, is to see all of these ministers of the Gospel. It's a marvelous picture. Who is it that is supposed to articulate the longings and aspirations of the people more than the preacher? Somehow the preacher must be an Amos, and say, "Let justice roll down like waters and righteousness like a mighty stream." Somehow, the preacher must say with Jesus, "The spirit of the Lord is upon me, because he hath anointed me to deal with the problems of the poor."

And I want to commend the preachers, under the leadership of these noble men: James Lawson, one who has been in this struggle for many years; he's been to jail for struggling; but he's still going on, fighting for the rights of his people. Rev. Ralph Jackson, Billy Kiles; I could just go right on down the list, but time will not permit. But I want to thank them all. And I want you to thank them, because so often, preachers aren't concerned about anything but themselves. And I'm always happy to see a relevant ministry.

It's all right to talk about "long white robes over yonder," in all of its symbolism. But ultimately people want some suits and dresses and shoes to wear down here. It's all right to talk about "streets flowing with milk and honey," but God has commanded us to be concerned about the slums down here, and his children who can't eat three square meals a day. It's all right to talk about the new Jerusalem, but one day, God's preacher must talk about the New York, the new Atlanta, the new Philadelphia, the new Los Angeles, the new Memphis, Tennessee. This is what we have to do.

Now the other thing we'll have to do is this: Always anchor our external direct action with the power of economic withdrawal. Now, we are poor people, individually, we are poor when you compare us with white society in America. We are poor. Never stop and forget that collectively, that means all of us together, collectively we are richer than all the nations in the world, with the exception of nine. Did you ever think about that? After you leave the United States, Soviet Russia, Great Britain, West Germany, France, and I could name the others, the Negro collectively is richer than most nations of the world. We have an annual income of more than thirty billion dollars a year, which is more than all of the exports of the United States, and more than the national budget of Canada. Did you know that? That's power right there, if we know how to pool it.

We don't have to argue with anybody. We don't have to curse and go around acting bad with our words. We don't need any bricks and bottles, we don't need any Molotov cocktails, we just need to go around to these stores, and to these massive industries in our country, and say, "God sent us by here, to say to you that you're not treating his children right. And we've come by here to ask you to make the first item on your agenda—fair treatment, where God's children are concerned. Now, if you are not prepared to do that, we do have an agenda that we must follow. And our agenda calls for withdrawing economic support from you."

And so, as a result of this, we are asking you tonight, to go out and tell your neighbors not to buy Coca-Cola in Memphis. Go by and tell them not to buy Sealtest milk. Tell them not to buy—what is the other bread?—Wonder Bread. And what is the other bread company, Jesse? Tell them not to buy Hart's bread. As Jesse Jackson has said, up to now, only the garbage men have been feeling pain; now we must kind of redistribute the pain. We are choosing these companies because they haven't been fair in their hiring policies; and we are choosing them because they can begin the process of saying, they are going to support the needs and the rights of these men who are on strike. And then they can move on downtown and tell Mayor Loeb to do what is right.

But not only that, we've got to strengthen black institutions. I call upon you to take your money out of the banks downtown and deposit your money in Tri-State Bank—we want a "bank-in" movement in Memphis. So go by the savings and loan association. I'm not asking you something that we don't do ourselves at SCLC. Judge Hooks and others will tell you that we have an account here in the savings and loan association from the Southern Christian Leadership Conference. We're just telling you to follow what we're doing. Put your money there. You have six or seven black insurance companies in Memphis. Take out your insurance there. We want to have an "insurance-in."

Now these are some practical things we can do. We begin the process of building a greater economic base. And at the same time, we are putting pressure where it really hurts. I ask you to follow through here.

Now, let me say as I move to my conclusion that we've got to give ourselves to this struggle until the end. Nothing would be more tragic than to stop at this point, in Memphis. We've got to see it through. And when we have our march, you need to be there. Be concerned about your brother. You may not be on strike. But either we go up together, or we go down together.

Let us develop a kind of dangerous unselfishness. One day a man came to Jesus; and he wanted to raise some questions about some vital matters in life. At points, he wanted to trick Jesus, and show him that he knew a little more than Jesus knew, and through this, throw him off base. Now that question could have easily ended up in a philosophical and theological debate. But Jesus immediately pulled that question from mid-air, and placed it on a dangerous curve between Jerusalem and Jericho. And he talked about a certain man, who fell among thieves. You remember that a Levite and a priest passed by on the other side. They didn't stop to help him. And finally a man of another race came by. He got down from his beast, decided not to be compassionate by proxy. But with him, administered first aid, and helped the man in need. Jesus ended up saying, this was the good man, this was the great man, because he had the capacity to project the "I" into the "thou," and to be concerned about his brother. Now you know, we use our imagination a great deal to try to determine why the priest and the Levite didn't stop. At times we say they were busy going to church meetings—an ecclesiastical gathering—and they had to get on down to Jerusalem so they wouldn't be late for their meeting. At other times we would speculate that there was a religious law that "One who was engaged in religious ceremonials was not to touch a human body twenty-four hours before the ceremony." And every now and then we begin to wonder whether maybe they were not going down to Jerusalem, or down to Jericho, rather to organize a "Jericho Road Improvement Association." That's a possibility. Maybe they felt that it was better to deal with the problem from the casual root, rather than to get bogged down with an individual effort.

But I'm going to tell you what my imagination tells me. It's possible that these men were afraid. You see, the Jericho road is a dangerous road. I remember when Mrs. King and I were first in Jerusalem. We rented a car and drove from Jerusalem down to Jericho. And as soon as we got on that road, I said to my wife, "I can see why Jesus used this as a setting for his parable." It's a winding, meandering road. It's really conducive for ambushing. You start out in Jerusalem, which is about 1200 miles, or rather 1200 feet above sea level. And by the time you get down to Jericho, fifteen or twenty minutes later, you're about 2200 feet below sea level. That's a dangerous road. In the days of Jesus it came to be known as the "Bloody Pass." And you know, it's possible that the priest and the Levite looked over that man on the ground and wondered if the robbers were still around. Or it's possible that they felt that the man on the ground was merely faking. And he was acting like he had been robbed and hurt, in order to seize them over there, lure them there for quick and easy seizure. And so the first question that the Levite asked was, "If I stop to help this man, what will happen to me?" But then the Good Samaritan came by. And he reversed the question: "If I do not stop to help this man, what will happen to him?"

That's the question before you tonight. Not, "If I stop to help the sanitation workers, what will happen to all of the hours that I usually spend in my office every day and every week as a pastor?" The question is not, "If I stop to help this man in need, what will happen to me?" "If I do not stop to help the sanitation workers, what will happen to them?" That's the question.

Let us rise up tonight with a greater readiness. Let us stand with a greater determination. And let us move on in these powerful days, these days of challenge to make America what it ought to be. We have an opportunity to make America a better nation. And I want to thank God, once more, for allowing me to be here with you.

You know, several years ago, I was in New York City autographing the first book that I had written. And while sitting there autographing books, a demented black woman came up. The only question I heard from her was, "Are you Martin Luther King?"

And I was looking down writing, and I said yes. And the next minute I felt something beating on my chest. Before I knew it I had been stabbed by this demented woman. I was rushed to Harlem Hospital. It was a dark Saturday afternoon. And that blade had gone through, and the X-rays revealed that the tip of the blade was on the edge of my aorta, the main artery. And once that's punctured, you drown in your own blood—that's the end of you.

It came out in the *New York Times* the next morning, that if I had sneezed, I would have died. Well, about four days later, they allowed me, after the operation, after my chest had been opened, and the blade had been taken out, to move around in the wheel chair in the hospital. They allowed me to read some of the mail that came in, and from all over the states, and the world, kind letters came in. I read a few, but one of them I will never forget. I had received one from the President and the Vice-President. I've forgotten what those telegrams said. I'd received a visit and a letter from the Governor of New York, but I've forgotten what the letter said. But there was another letter that came from a little girl, a young girl who was a student at the White Plains High School. And I looked at that letter, and I'll never forget it. It said simply, "Dear Dr. King: I am a ninth-grade student at the White Plains High School." She said, "While it should not matter, I would like to mention that I am a white girl. I read in the paper of your misfortune, and of your suffering. And I read that if you had sneezed, you would have died. And I'm simply writing you to say that I'm so happy that you didn't sneeze."

And I want to say tonight, I want to say that I am happy that I didn't sneeze. Because if I had sneezed, I wouldn't have been around here in 1960, when students all over the South started sitting-in at lunch counters. And I knew that as they were sitting in, they were really standing up for the best in the American dream. And taking the whole nation back to those great wells of democracy which were dug deep by the Founding Fathers in the Declaration of Independence and the Constitution. If I had sneezed, I wouldn't have been around in 1962, when Negroes in Albany, Georgia, decided to straighten their backs up. And whenever men and women straighten their backs up, they are going somewhere, because a man can't ride your back unless it is bent. If I had sneezed, I wouldn't have been here in 1963, when the black people of Birmingham, Alabama, aroused the conscience of this nation, and brought into being the Civil Rights Bill. If I had sneezed, I wouldn't have had a chance later that year, in August, to try to tell America about a dream that I had had. If I had sneezed, I wouldn't have been down in Selma, Alabama, to see the great movement there. If I had sneezed, I wouldn't have been in Memphis to see the community rally around those brothers and sisters who are suffering. I'm so happy that I didn't sneeze.

And they were telling me, now it doesn't matter now. It really doesn't matter what happens now. I left Atlanta this morning, and as we got started on the plane, there were six of us, the pilot said over the public address system, "We are sorry for the delay, but we have Dr. Martin Luther King on the plane. And to be sure that all of the bags were checked, and to be sure that nothing would be wrong with the plane, we had to check out everything carefully. And we've had the plane protected and guarded all night."

And then I got into Memphis. And some began to say the threats, or talk about the threats that were out. What would happen to me from some of our sick white brothers?

Well, I don't know what will happen now. We've got some difficult days ahead. But it doesn't matter with me now. Because I've been to the mountaintop. And I don't mind. Like anybody, I would like to live a long life. Longevity has its place. But I'm not concerned about that now. I just want to do God's will. And He's allowed me to go up to the mountain. And I've looked over. And I've seen the promised land. I may not get there with you. But I want you to know tonight, that we, as a people, will get to the promised land. And I'm happy, tonight. I'm not worried about anything. I'm not fearing any man. Mine eyes have seen the glory of the coming of the Lord.

Flip Schulke, ed., *Martin Luther King, Jr: A Documentary . . . Montgomery to Memphis* (New York and London: Norton, 1976), 222–23. The "Judge Hooks" referred to on page 199 is the Reverend Dr. Benjamin Hooks, then a local justice in Memphis, now executive director of the NAACP.

Index